Sergey N. Shevtsov • Arkady N. Soloviev
Ivan A. Parinov • Alexander V. Cherpakov
Valery A. Chebanenko

Piezoelectric Actuators and Generators for Energy Harvesting

Research and Development

Springer

Sergey N. Shevtsov
Russian Academy of Sciences
South Scientific Center of the Russian Academy of Sciences
Rostov-on-Don, Russia

Ivan A. Parinov
I. I. Vorovich Mathematics, Mechanics and Computer Science Institute
Southern Federal University
Rostov-on-Don, Russia

Valery A. Chebanenko
Russian Academy of Sciences
South Scientific Center of the Russian Academy of Sciences
Rostov-on-Don, Russia

Arkady N. Soloviev
Don State Technical University
Rostov-on-Don, Russia

Alexander V. Cherpakov
I. I. Vorovich Mathematics, Mechanics and Computer Science Institute
Southern Federal University
Rostov-on-Don, Russia

ISSN 2520-8047 ISSN 2520-8055 (electronic)
Innovation and Discovery in Russian Science and Engineering
ISBN 978-3-030-09283-2 ISBN 978-3-319-75629-5 (eBook)
https://doi.org/10.1007/978-3-319-75629-5

Printed on acid-free paper

This Springer imprint is published by the registered company Springer International Publishing AG part of Springer Nature.
The registered company address is: Gewerbestrasse 11, 6330 Cham, Switzerland

Innovation and Discovery in Russian Science and Engineering

Series editors

Carlos Brebbia
Wessex Institute of Technology, Southampton, United Kingdom

Jerome J. Connor
Department of Civil & Environmental Engineering, Massachusetts Institute of Technology, Cambridge, Massachusetts, USA

More information about this series at http://www.springer.com/series/15790

Preface

A great problem for modern science and modern techniques is the receiving, transformation, and storage of energy obtained from the environment and generated by working mechanisms and moving objects. While there is a fairly extensive scientific literature on R&D and application of energy-harvesting devices, significant breakthroughs in this field of science, technique, and technology have not yet been achieved. At present, investigations are planned into optimal constructions that allow obtaining maximum output characteristics of piezoelectric devices by using the specific geometry of the goods and high physical and mechanical characteristics of piezoelectric materials and composites, as well as the development of promising experimental, theoretical, and numerical methods for studying these complex technical systems.

This book presents some achievements and results in this field, obtained by the so-called Rostov Scientific School on Ferro-piezoelectricity. Investigations of the three-component systems, based on lead zirconate titanate (PZT), started in the Rostov State University (now Southern Federal University) at the end of the 1960s. Almost immediately, intensive studies of four- and five-component solid solutions were undertaken, and thereafter studies of six-component systems, based on PZT at the end of the 1990s. Piezoelectric Ceramic Rostovskaya (PCR) is a well-known brand that originally presented itself as PZT-type ceramics. Rostov scientists have developed and manufactured more than 100 systems of PCR over time, based on PZT composition as on the base of other ferro-piezoelectric solid solutions. Many materials, composites, and devices were developed, researched, and manufactured by these scientists in Rostov-on-Don. They have published more 5,000 journal papers and books on these topics and have been granted more 200 Soviet, Russian, and international patents (see, for example, monographs [4–8, 20, 50, 64, 120, 130, 135–139], and the references therein).

This book includes some of the latest results obtained by the scientists of the South Scientific Center of the Russian Academy of Sciences, Southern Federal University, and Don State Technical University (Rostov-on-Don). It also presents new approaches to R&D in piezoelectric generators and actuators of different types based on the developed original constructions and modern research into theoretical, experimental, and numerical methods of physics, mechanics, and materials science. Improved technical solutions of the devices are presented, which demonstrate high output values of voltage and power, allowing application of these products in various areas of energy harvesting.

The book is divided into seven chapters.

Chapter 1 considers a general overview of the problems of electro-elasticity in application to the investigation of energy harvesting, more specifically to the study of piezoelectric generators (PEGs). This chapter discusses constitutive equations of electro-elasticity in tensor form, and states corresponding boundary-value problems. Mathematical modelling of cantilever and stack types of piezoelectric generators are presented in detail. In particular, we consider bimorph piezoelectric structures with whole and partial covering of substrate by piezoelements. Numerous numerical results are presented for a broad spectrum of characteristics (in particular, first resonance frequency, voltage, and output power).

Chapter 2 discusses the developed original set-ups for testing the above-mentioned harvesters, samples of piezoelectric generators, and also the corresponding experimental methods and original computer algorithms. Comparisons of the obtained analytical and finite-element results with the experimental data obtained by using the developed test set-ups, are presented and discussed with the goal of optimizing construction of piezoelectric generators of both types. Experimental, numerical, and comparative results are obtained for cases of different kinds of loading (harmonical, pulsed, and quasi-statical).

Chapter 3 is devoted to mathematical modeling of the flexoelectric effect, arising in unpolarized piezoceramics under mechanical (in particular, bending) loading. There is discussion of the developed original set-up for estimation of this effect and the obtained experimental results for flexo-electrical beam under three-point bending. We formulate a corresponding boundary-value problem and obtain a theoretical solution that allows us to perform numerical experiments. The results allow studying the possibility of obtaining an electrical response, caused by the flexoelectric effect in ferroelectric ceramic plates of a certain composition. The numerical results show the possibility of the appearance of an electric potential in an unpolarized piezoceramic beam and also allows us to make conclusions on qualitative constituents of the theoretical model with the experiment.

Chapter 4 deals with the analytical and numerical modeling of the power of a high-stroke flex-tensional piezoelectric actuator, which consists of a high-power piezoelectric stack and polymeric composite shell, intended for amplification of the stroke. In order to overcome the principal drawback of the piezoelectric transducers, which is a very small stroke at relatively high operating force, an optimization problem is formulated and solved for the actuator's construction. For simultaneous provision of sufficient stroke and stiffness, allowing counteraction of the external loads, the shape of the amplified shell is parameterized by the rational Bezier curves. Their parameters (coordinates and weights of the control points) are changed iteratively by a genetic algorithm according to the objective function value, which is calculated by the finite element model of the transducer through varied geometry of the shell.

Since damage and defects have a crucial influence on all possible characteristics of the considered piezoelectric harvesters, the second part of the book is devoted to experimental-theoretical methods, computer simulation, and devices developed for the study and identification of defects in cantilever elastic rod constructions. Chapter 5 presents the current background for our studies in this area.

Chapter 6 is devoted to the development of methods for identifying the parameters of defects in an elastic cantilever with a notch, and the oscillation parameters are investigated in the context of dependence on the type of defect. The finite-element calculation of the modal parameters of full-body models of a cantilever rod with defect using the finite-element method is performed and the oscillation forms of the model are presented. The dependencies of natural frequencies on the defect location and size are investigated. The most sensitive modes of oscillations are determined with relation to dependence on the defect size at its different locations. The calculation of the dependence between the defect (notch) size of the cantilever rod of the full-body finite element model and the flexural rigidity of the elastic element is performed for the analytical model on the base of the dynamic equivalence of models.

Chapter 7 presents the measuring set-up that allows one to conduct technical diagnostics of rod constructions. It is based on the methods developed in the previous chapter. In addition, the results of the development and implementation of the algorithm of the calculation-experimental approach for the identification of defects in elements of cantilever structures are discussed. For this purpose, the original software and a laboratory information-measuring set-up have been developed, which provide an automated collection of information on construction vibrations and perform diagnostics of the defects.

The authors of the book especially thank V. A. Akopyan and E. V. Rozhkov for participating in the development of experimental approaches and creating test setups. We also acknowledge the Russian Foundation for Basic Research and Russian Ministry of Education and Science, grants from which helped to perform this research.

This self-standing book, covering the necessary theoretical, experimental, and numerical modeling approaches, is aimed at a wide range of students, engineers, and specialists interested and participating in R&D of modern energy-harvesting devices, the materials for these devices, the development of physical and mathematical methods for their study, and also experimental equipment for definition of their characteristics.

Rostov-on-Don, Russia
December, 2017

Sergey N. Shevtsov
Arkady N. Soloviev
Ivan A. Parinov
Alexander V. Cherpakov
Valery A. Chebanenko

Contents

Chapter 1
Mathematical Modeling of Piezoelectric Generators

In recent years, the research on piezoelectric transducers of mechanical energy into electrical energy has been actively developed. This type of transducer is called a piezoelectric generator (PEG). Basic information about PEGs, as well as the problems arising at the development stages of energy harvesting devices, was provided in the review papers [97, 95, 38, 73], as well as in the monographs [21, 60, 62]. PEGs are manufactured in two different configurations: stack and cantilever. PEGs have a broad application, e.g., for piezoelectric damping of vibrations [13, 15, 145, 143].

Most researchers have studied the characteristics of the cantilever-type PEGs. There are several ways of modeling PEGs: (i) mathematical model with lumped parameters, (ii) mathematical model with distributed parameters, and (iii) finite element model.

Works [22, 59, 58, 3, 143, 144] are devoted to the construction of PEG models based on oscillations of a mechanical system with lumped parameters. The use of such systems is a convenient model approach, since researchers can obtain analytical dependencies between the output parameters of PEG (voltage, power, etc.) and electro-mechanical characteristics, and also electrical resistance of the external electric circuit.

The modeling with lumped parameters provides initial representations on the problem, allowing one to use simple expressions for the description of the system. However, it is approximate and restricted by only one oscillation mode. This description does not take into account other important aspects of the system.

Another type of modeling is distributed parameter modeling. Based on the Euler-Bernoulli hypothesis for beams, analytical solutions of the coupled problem have been obtained in [61, 53, 158, 157] for different configurations of cantilever-type PEGs. The authors obtained clear expressions for the output voltage on resistive electric load and for console displacements. Moreover, they studied, in detail, behaviors of PEGs with short-circuited and open-circuited electric circuits, and the influence of the piezoelectric couple effect and flexoelectric effect [53, 158].

S. N. Shevtsov et al., *Piezoelectric Actuators and Generators for Energy Harvesting*, Innovation and Discovery in Russian Science and Engineering,
https://doi.org/10.1007/978-3-319-75629-5_1

Nevertheless, in these studies, the case when the piezoelectric element does not completely cover the substrate has not been considered.

Works [154, 123, 162, 164, 182, 161] are devoted to the finite element modeling of the different types of cantilever PEG. The case when the piezoelectric element does not completely cover the substrate is easily solved using this modeling approach. Nevertheless, obtaining a semi-analytical solution for incomplete covering of the substrate by a piezoelectric element is of interest.

Most works devoted to the investigation of stack-type PEGs are based on finite element modeling [160, 33, 35, 63] and lumped parameter modeling [59, 69, 184]. Recently, attention has been directed to analytical studies of stack-type generators. Due to the ability of stack PEGs to carry high compression levels that allow their integration in different infrastructure objects (e.g., transport roads and rail-roads), there is a need to develop mathematical models for predicting PEGs output characteristics.

Several models of stack-type PEGs have been proposed in [184, 180]. The model discussed in [184] depends on the initial experimental data and does not provide information about displacements. The model proposed in [180] does not have such shortcomings. However, it is very tedious for analysis due to its recursive type.

The above brief analysis of known works has shown that the problem of modeling different types of PEGs with the help of analytical methods has not been solved, yet is quite relevant.

1.1 General Formulation of the Problem of Electroelasticity

The basic equations in the theory of electroelasticity are the equations of motion and the equations of the electric field [176]:

$$\begin{aligned} \sigma_{ji,j} + X_i &= \rho \ddot{u}_i \\ D_{i,i} &= 0, \ \ \mathbf{x} \in V, t > 0 \end{aligned} \tag{1.1}$$

where σ_{ij} are the components of the stress tensor, X_i are the components of the vector of mass forces, u_i are the components of the displacement vector, D_i are the components of the electric displacement vector, V is the body volume.

To these equations, the constitutive relations for the electroelastic body are added [176]:

$$\begin{aligned} \sigma_{ij} &= c^E_{ijkl} \varepsilon_{kl} - e_{kij} E_k \\ D_i &= e_{ikl} \varepsilon_{kl} + \text{э}^S_{ik} E_k \end{aligned} \tag{1.2}$$

where c^E_{ijkl} are the components of the elastic moduli tensor, measured at a constant electric field, ε_{kl} are the components of the linear deformation tensor, e_{kij} are the components of the piezoelectric constants tensor, E_k are the components of the electric field vector, $\ni^{s}_{ik}$ are the components of dielectric constants tensor, measured at a constant displacement.We will consider a quasi-static electric field and the linear deformation:

$$\begin{aligned} \varepsilon_{ij} &= \frac{1}{2}\left(u_{i,j} + u_{j,i}\right) \\ E_i &= -\varphi_{,i} \end{aligned} \tag{1.3}$$

where ϕ is the electric potential

By substituting (1.2) and (1.3) into (1.1), we obtain a system of coupled equations in which the unknowns are the displacement u_i and the electric potential ϕ:

$$\begin{aligned} c^{E}_{ijkl}u_{k,lj} + e_{kij}\varphi_{,kj} + X_i &= \rho\ddot{u}_i \\ e_{ikl}u_{k,li} - \ni^{s}_{ik}\varphi_{,ki} &= 0 \end{aligned} \tag{1.4}$$

The first equation describes the motion and the second describes the quasi-static electric field.

We add boundary conditions to these equations. Let the surface S consist of two parts Γ_1 and Γ_2 so $S = \Gamma_1 \cup \Gamma_2$, where $\Gamma_1 \cap \Gamma_2 = 0$. Suppose that the displacement U_i are given on Γ_1 and the loads p_i are given on Γ_2. Then the boundary conditions will have the following form:

$$\begin{aligned} u_i\big|_{\Gamma_1} &= U_i(\mathbf{x}, t), \quad \mathbf{x}\in\Gamma_1, \\ \sigma_{ji}n_j\big|_{\Gamma_2} &= p_i(\mathbf{x}, t), \quad \mathbf{x}\in\Gamma_2. \end{aligned} \tag{1.5}$$

Again, we divide the surface S into two parts Γ_3 and Γ_4, moreover $S = \Gamma_3 \cup \Gamma_4$ and $\Gamma_3 \cap \Gamma_4 = 0$. Let the electric potential ϕ is given on the surface $\Gamma_3 = \cup_{k=1}^{M}\Gamma_3^k,\ M\in Z$, and the surface charge σ_0 is given on the Γ_4. Then we obtain the boundary conditions of the following form:

$$\begin{aligned} \varphi\big|_{\Gamma_3^k} &= v_k(t), \mathbf{x}\in\Gamma_3 \\ D_k n_k\big|_{\Gamma_4} &= -\sigma_0(\mathbf{x}, t), \mathbf{x}\in\Gamma_4 \end{aligned} \tag{1.6}$$

It remains to add the initial conditions for displacements:

$$u_i(\mathbf{x}, 0) = f_i(\mathbf{x}), \quad \dot{u}_i(\mathbf{x}, 0) = g_i(\mathbf{x}), \quad \mathbf{x}\in V, \quad t = 0. \tag{1.7}$$

In the case when the k-th electrode is connected to an external circuit, the following condition must be added:

$$\iint\limits_{\Gamma_3^k} \dot{D}_i n_i ds = I. \tag{1.8}$$

With help of this condition, the unknown electric potential v_k can be found using the electric current I.

1.2 Mathematical Modelling of Cantilever-type PEGs

1.2.1 Statement of the Problem for Cantilever-type PEGs

Let us consider the functional [163]:

$$\Pi = \iiint\limits_V (H - X_i u_i) dV - \iint\limits_S (p_i u_i + \sigma\varphi) dS, \tag{1.9}$$

where H is the electric enthalpy. The Hamilton principle, generalized to the theory of electroelasticity, has the form:

$$\delta \int\limits_{t_1}^{t_2} (K - \Pi) dt = 0 \tag{1.10}$$

where K is the kinetic energy.

Substituting (1.9) into (1.10), we obtain the following expression for the Hamilton principle:

$$\int\limits_{t_1}^{t_2} dt \iiint\limits_V (\delta K - \delta H) dV + \int\limits_{t_1}^{t_2} dt \left[\iiint\limits_V X_i \delta u_i dV + \iint\limits_S (p_i \delta u_i + \sigma \delta\varphi) dS \right] = 0. \tag{1.11}$$

The variation of the electric enthalpy in linear electroelasticity is:

$$\delta H = \sigma_{ij} \delta\varepsilon_{ij} - D_i \delta E_i. \tag{1.12}$$

The variation of the kinetic energy is:

$$\delta \int\limits_{t_1}^{t_2} K dt = -\rho \int\limits_{t_1}^{t_2} dt \iiint\limits_V \ddot{u}_i \delta u_i dV. \tag{1.13}$$

Next, we will consider the case of the absence of mass forces and external loads. Surface charge densities are assumed to be unknown.

Let us consider the simplest bimorph design of cantilever-type PEGs, presented in Fig. 1.1. Basic configuration of a cantilever bimorph PEG is two piezoelements glued to the substrate. This construction is clamped at one end and the other end remains free. The piezoelectric elements are connected in parallel and connected to an external circuit, consisting of a resistor R. The potential difference $v(t)$ will be measured on this resistor. The thickness of the electrodes and the adhesive layer, due to the smallness of their values, can be neglected.

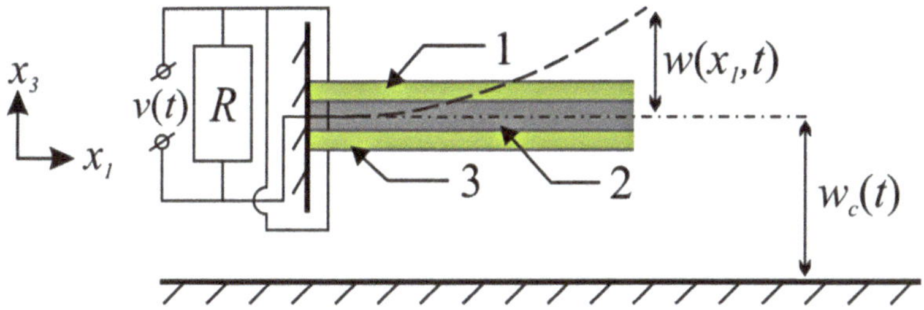

Fig. 1.1 Bimorph cantilever PEG: 1 and 3 – piezoelements, 2 – substrate

To simplify the description of the behavior of the structure shown in Fig. 1.1, we introduce the Euler-Bernoulli hypothesis. The excitation of oscillations in the PEG occurs through the movement of the clamp against a certain plane. Therefore, the absolute displacement of the cantilever along the coordinate x_3 will consist of the sum of the movements of the clamp $w_c(t)$ and the relative displacement of the cantilever $w(x_1, t)$. Taking into account the above, the vector of displacement $\mathbf{u}$ takes the following form:

$$\mathbf{u} = \left\{ -x_3 \frac{\partial w(x_1, t)}{\partial x_1}, 0, w(x_1, t) + w_c(t) \right\}^T. \tag{1.14}$$

By introducing the Euler-Bernoulli hypothesis, we consider the one-dimensional problem. This simplifies the constitutive relations (1.2):

$$\begin{aligned} \sigma_{11} &= c_{11}^{E*} \varepsilon_{11} - e_{31}{}^* E_3 \\ D_3 &= e_{31}{}^* \varepsilon_{11} + \ni_{33}^{S*} E_3 \end{aligned} \tag{1.15}$$

where the material constants are expressed as follows:

$$c_{11}^{E*} = \frac{1}{s_{11}^E}, \; e_{31}{}^* = \frac{d_{31}}{s_{11}^E}, \; \ni_{33}^{S*} = \ni_{33}^T - \frac{d_{31}^2}{s_{11}^E}. \tag{1.16}$$

In the PEG studied, the electrodes on piezoelements are applied to large sides perpendicular to the axis, and therefore it makes sense to consider only the components of the electric potential along the axis x_3.

It is assumed that the piezoelectric element is sufficiently thin, and there are no free charges inside it. Therefore, we introduce the assumption of a linear electric field distribution over the thickness of the piezoceramic element:

$$\varphi_{,3} = \frac{v(t)}{h_p} \tag{1.17}$$

where $v(t)$ is the potential difference between the upper and lower electrode of the piezoelectric element, h_p is the thickness of the piezoelectric element.

Taking into account all the assumptions and hypotheses introduced, the Hamilton principle (1.11) takes the following form:

$$\int_{t_1}^{t_2} dt \left\{ \iiint_V \left[\left(-c_{11}^{E*} x_3^2 \frac{\partial^2 w(x_1,t)}{\partial {x_1}^2} + e_{31}{}^* x_3 \frac{v(t)}{h_p} \right) \delta \left(\frac{\partial^2 w(x_1,t)}{\partial {x_1}^2} \right) \right. \right.$$
$$+ \left(\frac{e_{31}{}^* x_3}{h_p} \frac{\partial^2 w(x_1,t)}{\partial {x_1}^2} + \ni_{33}^{S*} \frac{v(t)}{h_p^2} \right) \delta(v(t)) \tag{1.18}$$
$$\left. -\rho(\ddot{w}(x_1,t) - \ddot{w}_c(t))\delta w(x_1,t) \right] dV + \iint_S \frac{\sigma x_3}{h_p} \delta v(t) dS \bigg\} = 0$$

To analyze this problem, it is convenient to use the semi-discrete Kantorovich method [81]. To do this, we represent the relative displacements of a beam as an expansion in a series:

$$w(x_1,t) = \sum_{i=1}^{N} \eta_i(t) \phi_i(x_1) \tag{1.19}$$

where N is the number of vibration modes considered, $\eta_i(t)$ are the unknown generalized coordinates, $\varphi_i(x_1)$ are the known trial functions that satisfy the boundary conditions.

After substituting the expansion (1.19) into the expression for the Hamilton principle (1.18), we zero the coefficients for independent variations of $\delta\eta$ and δv. The integration of the charge density σ over the area S gives us an unknown charge q:

$$\iint_S \sigma dS = q. \tag{1.20}$$

Thus, we obtain a system of differential equations. In order to take into account the influence of the external electric circuit, we differentiate the second equation in this system by taking into account the fact that $\dot{q} = I$. Further, using Ohm's law, we obtain a system of differential equations, describing the forced oscillations of a bimorph PEG connected to a resistor:

$$\begin{aligned} &\mathbf{M}\ddot{\boldsymbol{\eta}}(t) + \mathbf{D}\dot{\boldsymbol{\eta}}(t) + \mathbf{K}\boldsymbol{\eta}(t) - \boldsymbol{\Theta}v(t) = \mathbf{p} \\ &C_p \dot{v}(t) + \boldsymbol{\Theta}^{\mathrm{T}} \dot{\boldsymbol{\eta}}(t) + \frac{v(t)}{R} = 0 \end{aligned} \tag{1.21}$$

where $\mathbf{D} = \mu\mathbf{M} + \gamma\mathbf{K}$ is the Rayleigh-type damping matrix, and the remaining coefficients are:

$$C_p = \frac{b_p L_p}{h_p} \ni_{33}^{S} *$$

$$M_{ij} = \int_0^L m\varphi_i(x_1)\varphi_j(x_1)dx_1$$

$$K_{ij} = \int_0^L \mathrm{EI}\varphi_i''(x_1)\varphi_j''(x_1)dx_1 \tag{1.22}$$

$$p_i = -\ddot{w}_c(t)\left[\int_0^L m\varphi_i(x_1)dx_1\right]$$

$$\theta_i = \int_0^L J_p\varphi_i''(x_1)dx_1$$

where C_p is the electric capacitance, b_p is the width of the piezoelectric element, L_p is the length of the piezoelement, M_{ij} are the elements of the mass matrix, K_{ij} are the elements of the stiffness matrix, θ_i are the elements of the electromechanical coupling vector, p_i are the elements of the effective mechanical load vector, m is the specific mass, EI is the flexural stiffness.

Let us move on to solving the system of Eq. (1.21). We assume that the excitation $w_c(t)$ is harmonic:

$$\begin{aligned} w_c(t) &= \tilde{w}_c e^{i\omega t} \\ \mathbf{p} &= \tilde{\mathbf{p}} e^{i\omega t} \end{aligned} \tag{1.23}$$

then the solution is sought in the form:

$$\begin{aligned} \boldsymbol{\eta}(t) &= \tilde{\boldsymbol{\eta}} e^{i\omega t} \\ \nu(t) &= \tilde{\nu}(t) e^{i\omega t} \end{aligned} \tag{1.24}$$

A wave over a variable in the Eqs. (1.23) and (1.24) indicates the amplitude. After substituting (1.23) and (1.24), the system of Eq. (1.21) takes the form:

$$\begin{aligned} &[-\omega^2\mathbf{M} + i\omega(\mu\mathbf{M} + \gamma\mathbf{K}) + \mathbf{K}]\tilde{\boldsymbol{\eta}} - \boldsymbol{\Theta}\tilde{\nu} = \tilde{\mathbf{p}} \\ &\left(i\omega C_p + \frac{1}{R}\right)\tilde{\nu} + i\omega\boldsymbol{\Theta}^T\tilde{\boldsymbol{\eta}} = 0 \end{aligned} \tag{1.25}$$

After some simple algebraic transformations, we obtain solutions for $\tilde{\boldsymbol{\eta}}$ and $\tilde{\nu}$ in the case of harmonic base excitation:

$$\tilde{\boldsymbol{\eta}} = \left[-\omega^2 \mathbf{M} + i\omega(\mu\mathbf{M} + \gamma\mathbf{K}) + \mathbf{K} + \frac{i\omega \boldsymbol{\Theta}\boldsymbol{\Theta}^T}{i\omega C_p + \frac{1}{R}} \right]^{-1} \tilde{\mathbf{p}}$$
$$\tilde{v} = -\frac{i\omega \boldsymbol{\Theta}^T}{i\omega C_p + \frac{1}{R}} \left[-\omega^2 \mathbf{M} + i\omega(\mu\mathbf{M} + \gamma\mathbf{K}) + \mathbf{K} + \frac{i\omega \boldsymbol{\Theta}\boldsymbol{\Theta}^T}{i\omega C_p + \frac{1}{R}} \right]^{-1} \tilde{\mathbf{p}} \tag{1.26}$$

1.2.2 Numerical Experiment

The Kantorovich method in Eq. (1.19), used for analyzing the model, is associated with the application of trial functions $\phi_i(x_1)$. Since we expand the displacement $w(x_1, t)$ in a series, the trial functions must satisfy the mechanical boundary conditions. In order to find these functions, it is necessary to solve the eigenvalue problem for the beam.

Next, four configurations of cantilever-type PEGs will be considered, with each configuration having their own design features. To take into account these features, the beam is divided into longitudinal segments. Therefore, in addition to the boundary conditions, we also have to use the coupling conditions for the beam segments.

For further calculations, unless otherwise specified, the physical and geometric properties given in Table 1.1 will be used.

The excitation in the system is given by the harmonic displacement of the base $w_c = \tilde{w}_c e^{i\omega t}$, whose amplitude is 0.1 mm, and the coefficients of the modal damping of the damping are the same: $\xi_1 = \xi_2 = 0.02$.

From this point forward, the output electric power is calculated by the formula:

$$P = \frac{v^2}{R} \tag{1.27}$$

Table 1.1 PEG Parameters

	Substrate	Piezoelement
Geometrical dimensions ($L_0 \times b \times h$)	$110 \times 10 \times 1$ mm^3	$56 \times 6 \times 0.5$ mm^3
Density (ρ)	1,650 kg/m^3	8,000 kg/m^3
The Young's modulus and Poisson's ratio (E, ν)	15 GPa and 0.12	–
Elastic compliance (s_{11}^E)	–	17.5×10^{-12} Pa
Relative permittivity ($\varepsilon_{33}^S/\varepsilon_0$)	–	5,000
Piezoelectric module (d_{31})	–	−350 pC/N

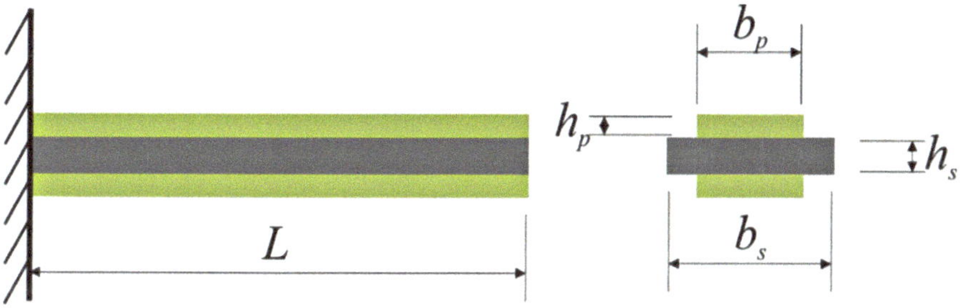

Fig. 1.2 Bimorph cantilever PEG and its cross-section

1.2.2.1 Configuration # 1

Let us start with the simplest case, shown in Fig. 1.2, when the length of the piezoelements is the same as the length of the substrate. The variables h_s and b_s in Fig. 1.2 denote the height and width of the substrate, respectively, and h_p and b_p denote the height and width of the piezoelectric element, respectively. Further in the text, the subscripts s and p in variables will indicate the correspondence of these variables to the substrate or piezoelement, respectively. To find $\varphi_i(x_1)$ we need solve the problem of free oscillations of a given beam.

Since this configuration is quite simple, it has been studied in many works of other authors by various methods. We will not give numerical results for it, but only a general solution.

We write out the solution of the oscillation equation in the common form:

$$\begin{aligned}\phi_i(x_1) = a_{1,i}\sin(\beta_i x_1) + a_{2,i}\cos(\beta_i x_1) + a_{3,i}\sinh(\beta_i x_1)\\ + a_{4,i}\cosh(\beta_i x_1).\end{aligned} \tag{1.28}$$

We write the boundary conditions:

$$\begin{aligned}\phi_i(0) = 0\ \phi_i''(L) = 0\\ \phi_i'(0) = 0\ \phi_i'''(L) = 0\end{aligned} \tag{1.29}$$

Satisfying the boundary conditions, we obtain a homogeneous system of four equations with four unknowns. We write it in the matrix form:

$$\Lambda = \begin{pmatrix} a_{1,1} & \cdots & a_{1,4} \\ \vdots & \ddots & \vdots \\ a_{4,1} & \cdots & a_{4,4} \end{pmatrix} = 0. \tag{1.30}$$

In order for a given system to have non-zero solutions, it is necessary and sufficient that its determinant be zero. Having found the determinant of the matrix (1.30), we obtain the characteristic equation from which we find the eigenvalues β_i:

$$1 + \cos\beta_i \cosh\beta_i = 0. \tag{1.31}$$

Since Eq. (1.31) is a transcendental equation, we will seek its solution using numerical methods. After receiving the set of β_i, we can calculate the coefficients a_i for the required number of modes N.

In this configuration, the specific mass, which appears in Eq. (1.22), for the section shown in Fig. 1.2, is calculated as follows:

$$m = \rho_s A_s + 2\rho_p A_p, \tag{1.32}$$

where $A = hb$ is the cross-sectional area.

The bending stiffness EI for this configuration is calculated as follows:

$$\mathrm{EI} = c_p \left[\iint\limits_{S_{pu}} x_3^2 dS + \iint\limits_{S_{pl}} x_3^2 dS \right] + c_s \iint\limits_{S_s} x_3^2 dS, \tag{1.33}$$

where c_p and c_s denote the corresponding elastic moduli, respectively, and S_{pl} and S_{pu} are the cross-section areas along which integration is performed for the lower and upper piezoelements, respectively.

The function J_p is defined as

$$J_p = \frac{e_{31}{}^*}{h_p} \left(\iint\limits_{S_{pl}} x_3 dS + \iint\limits_{S_{pu}} x_3 dS \right) \tag{1.34}$$

The case considered above can be modified by adding the proof point mass M to the free end of the beam (see Fig. 1.3).

The boundary conditions at the free end of the beam in this case will change as follows:

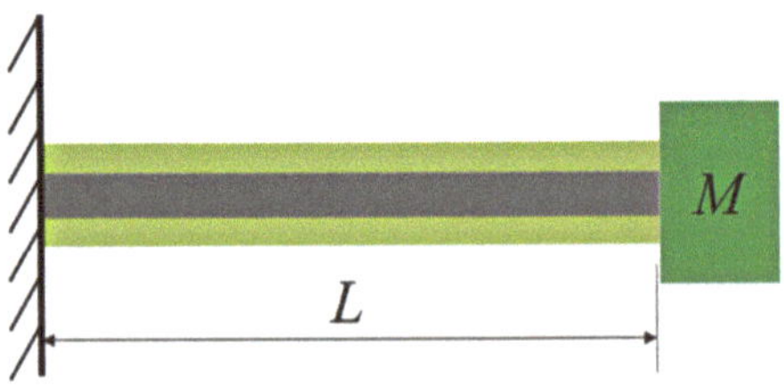

Fig. 1.3 Bimorph cantilever PEG with proof mass

$$\begin{aligned}
&\phi_i''(L) = 0 \\
&\phi_i'''(L) = -\alpha\beta^4\varphi_i(L) \\
&\alpha = \frac{M}{mL}
\end{aligned} \tag{1.35}$$

The characteristic equation in this case takes the following form:

$$1 + \cos(\beta_i)\cosh(\beta_i) - \frac{\sin(\beta_i)\cosh(\beta_i)\beta_i m}{ML} + \frac{\cos(\beta_i)\sinh(\beta_i)\beta_i m}{ML} = 0. \tag{1.36}$$

Adding a proof mass to the construction requires considering its effect on the system of Eq. (1.21), since it is an additional inertial load that affects the kinetic energy. Taking into account the proof mass, the expressions for some components in Eq. (1.22) change as follows:

$$\begin{aligned}
M_{ij} &= \int_0^L m\varphi_i(x_1)\phi_j(x_1)dx_1 + M\phi_i(L_M)\phi_j(L_M) \\
p_i &= -\ddot{w}_c(t)\int_0^L m\phi_i(x_1)dx_1 + M\phi_i(L_M)
\end{aligned} \tag{1.37}$$

where L_M is the coordinate of the proof mass. In this case, it is the free end of the beam.

1.2.2.2 Configuration # 2

Now we consider the more complicated case shown in Fig. 1.4, which requires dividing the beam into two segments.

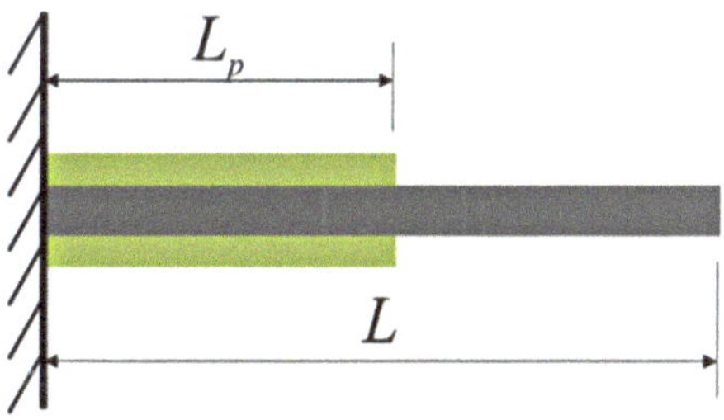

Fig. 1.4 Bimorph cantilever PEG with incomplete substrate coating by piezoelectric elements

To solve the problem of finding eigenvalues for such a construction, we define the function $\phi_i(x_1)$ as a piecewise-defined function:

$$\varphi_i(x_1) = \begin{cases} \varphi_i^{(1)}(x_1), x_1 \leq L_p \\ \varphi_i^{(2)}(x_1), x_1 > L_p \end{cases} \tag{1.38}$$

where $\phi_i^{(1)}$ corresponds to the shape of the oscillations of the left segment of the beam covered by the piezoelectric element, and $\phi_i^{(2)}$ corresponds to the right segment, L_p is the length of the piezoelectric element. We write the solution in common form for each part of the beam:

$$\begin{aligned} \phi_i^{(1)}(x_1) &= a_{1,i}\sin(\beta_i x_1) + a_{2,i}\cos(\beta_i x_1) + a_{3,i}\sinh(\beta_i x_1) + a_{4,i}\cosh(\beta_i x_1) \\ \phi_i^{(2)}(x_1) &= a_{5,i}\sin(\beta_i x_1) + a_{6,i}\cos(\beta_i x_1) + a_{7,i}\sinh(\beta_i x_1) + a_{8,i}\cosh(\beta_i x_1) \end{aligned} \tag{1.39}$$

Let us write down the boundary conditions for the ends of the beam and the conjugation condition at that point of the beam where the piezoelectric element ends:

$$\begin{array}{lcl} & \phi_i^{(1)}(L_p) = \phi_i^{(2)}(L_p) & \\ & \phi_i^{(1)'}(L_p) = \phi_i^{(2)'}(L_p) & \\ \phi_i^{(1)}(0) = 0, & & \phi_i^{(2)''}(L_p) = 0 \\ \phi_i^{(1)'}(0) = 0, & \phi_i^{(1)''}(L_p) = \dfrac{\mathrm{EI}^{(2)}}{\mathrm{EI}^{(1)}}\phi_i^{(2)''}(L_p) & \phi_i^{(2)''}(L_p) = 0 \\ & \phi_i^{(1)'''}(L_p) = \dfrac{\mathrm{EI}^{(2)}}{\mathrm{EI}^{(1)}}\phi_i^{(2)'''}(L_p) & \end{array} \tag{1.40}$$

where $\mathrm{EI}^{(1)}$ is the flexural stiffness of the left segment, and $\mathrm{EI}^{(2)}$ corresponds to the right segment.

Satisfying the boundary conditions, we obtain a homogeneous system of eight equations with eight unknowns:

$$\Lambda = \begin{pmatrix} a_{1,1} & \cdots & a_{1,8} \\ \vdots & \ddots & \vdots \\ a_{8,1} & \cdots & a_{8,8} \end{pmatrix}. \tag{1.41}$$

As in the previous case, we find the determinant of the given system and solve the characteristic equation by numerical methods. Because of the cumbersomeness, here and below, the characteristic equations will not be cited.

The specific mass m, for the construction, shown in Fig. 1.4, is calculated as follows:

$$m(x_1) = \rho_s A_s + 2\rho_p A_p G(x_1), \tag{1.42}$$

where $G(x_1)$ is the function responsible for the position of the piezoelectric element on the substrate. In this configuration, it equals:

$$G(x_1) = 1 - H(x_1 - L_p), \tag{1.43}$$

where $H(x_1)$ is the Heaviside function.

The flexural rigidity $\mathrm{EI}(x_1)$ for this model is calculated in a similar way:

$$\mathrm{EI}(x_1) = c_p \left[\iint\limits_{S_{pl}} x_3^2 dS + \iint\limits_{S_{pu}} x_3^2 dS \right] G(x_1) + c_{\Pi} \iint\limits_{S_s} x_3^2 dS. \tag{1.44}$$

The function $J_p(x_1)$ is equal to:

$$J_p(x_1) = \frac{e_{31}{}^*}{h_p} \left(\iint\limits_{S_{pu}} x_3 dS + \iint\limits_{S_{pl}} x_3 dS \right) G(x_1). \tag{1.45}$$

The solution for the model shown in Fig. 1.5 is exactly the same as for the model in Fig. 1.4.

The main difference is in the coupling conditions in the place of attachment of the mass:

$$\begin{aligned}
&\phi_i^{(1)}(L_M) = \phi_i^{(2)}(L_M), \\
&\phi_i^{(1)\prime}(L_M) = \phi_i^{(2)\prime}(L_M), \\
&\phi_i^{(1)\prime\prime}(L_M) = \phi_i^{(2)\prime\prime}(L_M), \\
&\phi_i^{(1)\prime\prime\prime}(L_M) = \phi_i^{(2)\prime\prime\prime}(L_M) - \alpha\beta^4\phi_i^{(1)}(L_M), \\
&\alpha = \frac{M}{mL}
\end{aligned} \tag{1.46}$$

The specific mass m, bending stiffness EI, and function $J_p(x_1)$ are found from Eqs. (1.32), (1.33), and (1.34), respectively.

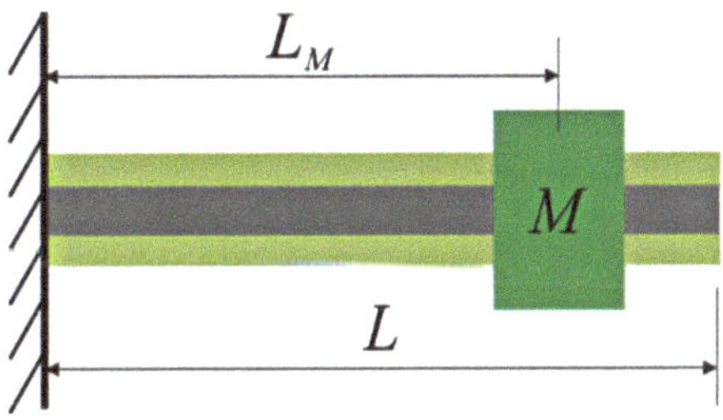

Fig. 1.5 Bimorph cantilever PEG with displaced proof mass

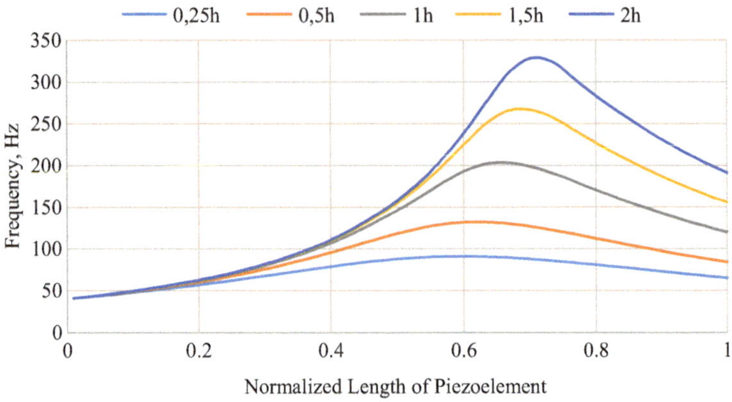

Fig. 1.6 First resonance frequency vs. normalized length of piezoelement for different thicknesses of the piezoelectric elements

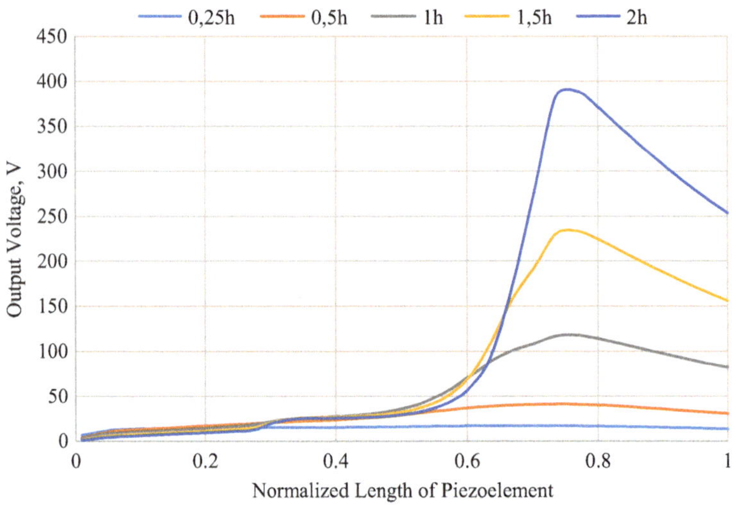

Fig. 1.7 Output voltage vs. normalized length of piezoelement for different thicknesses of the piezoelectric elements

Further, on the base of the obtained solution for the model shown in Fig. 1.4, we construct the dependencies of the basic performance characteristics of PEG as a function of the length of the piezoelectric element. For convenience in analyzing the results, the length of the piezoelements was normalized with respect to the length of the substrate. All the dependencies for cantilever-type PEGs were obtained for different thicknesses of the piezoelectric element h_p (h in figures below), which was calculated by multiplying the thickness of the substrate h_s by a certain coefficient (for example:$h_p = 0.5h$). All characteristics are investigated at the first resonance frequency and presented in Figs. 1.6, 1.7, 1.8 and 1.9.

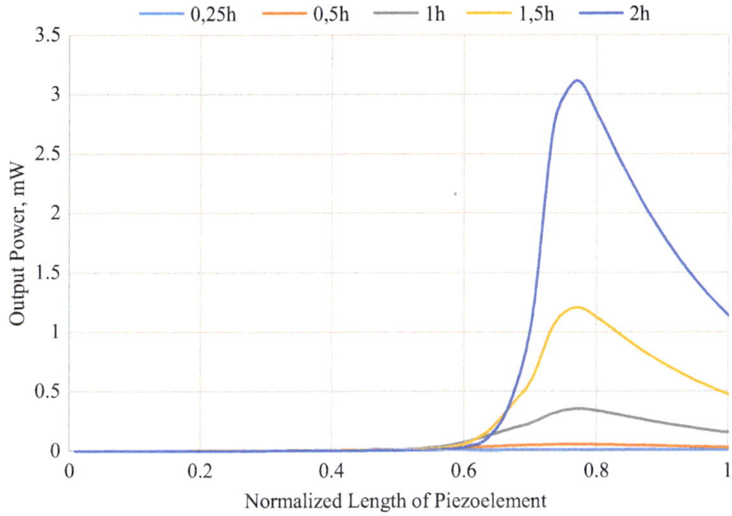

Fig. 1.8 Output power vs. normalized length of piezoelement for different thicknesses of the piezoelectric elements

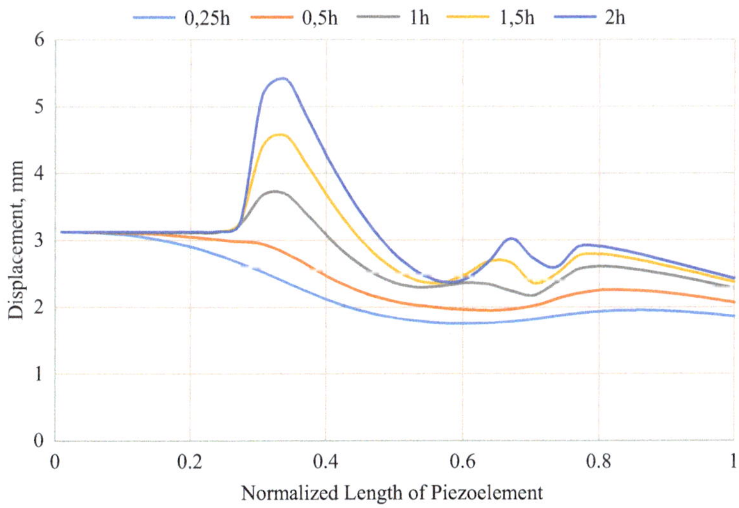

Fig. 1.9 Displacement of the beam's free end vs. normalized length of piezoelement for different thicknesses of the piezoelectric elements

Analysis of Figs. 1.6, 1.7, 1.8, and 1.9 allows us to draw the following conclusions:

(i) increasing the length of the piezoelectric element increases the resonance frequency, output voltage, and output power to some maximum values, after which the values of these characteristics begin to decrease;
(ii) increasing the thickness of the piezoelectric element also increases the resonance frequency, output voltage, and output power;
(iii) in the range of the investigated values, the maximum output power is reached when the piezoelectric element is twice as thick as the substrate and covers about 80 percent of its length;
(iv) the dependence of the displacement of the end of the beam on the length of the piezoelement for various thicknesses of the piezoelements has a complex shape.

1.2.2.3 Configuration # 3

Now we consider the more complicated case, shown in Fig. 1.10.

To find the solution for this construction, we divide the beam into three segments. The first segment includes from the clamping up to the beginning of the piezoelement. The second segment is a part of the substrate, covered with a piezoelectric element. The third segment is the rest of the beam after the piezoelement. We reflect this division of a beam in a piecewise-defined function $\phi_i(x_1)$ as:

$$\phi_i(x_1) = \begin{cases} \phi_i^{(1)}(x_1), x_1 \leq L_0 \\ \phi_i^{(2)}(x_1), L_0 < x_1 \leq L_p + L_0 \\ \phi_i^{(3)}(x_1), x_1 > L_p + L_0 \end{cases} \quad (1.47)$$

where $\phi_i^{(1)}$, $\phi_i^{(2)}$ and $\phi_i^{(3)}$ correspond to the oscillation forms of the first, second, and third segments, respectively, and L_0 is the length of the first segment. We write the solution in common form for each part of the beam:

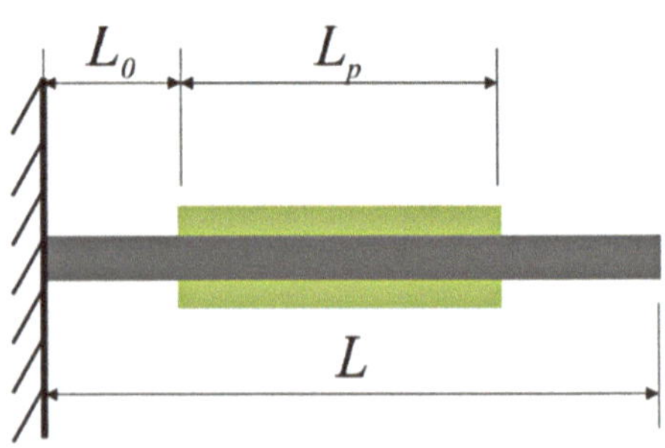

Fig. 1.10 Bimorph cantilever PEG with offset piezoelectric element relative to the clamp

$$\begin{aligned}
\phi_i^{(1)}(x_1) &= a_{1,i}\sin(\beta_i x_1) + a_{2,i}\cos(\beta_i x_1) + a_{3,i}\sinh(\beta_i x_1) + a_{4,i}\cosh(\beta_i x_1)\\
\phi_i^{(2)}(x_1) &= a_{5,i}\sin(\beta_i x_1) + a_{6,i}\cos(\beta_i x_1) + a_{7,i}\sinh(\beta_i x_1) + a_{8,i}\cosh(\beta_i x_1)\\
\phi_i^{(3)}(x_1) &= a_{9,i}\sin(\beta_i x_1) + a_{10,i}\cos(\beta_i x_1) + a_{11,i}\sinh(\beta_i x_1) + a_{12,i}\cosh(\beta_i x_1)
\end{aligned} \tag{1.48}$$

The boundary conditions for the ends of the beam remain the same as in Eq. (1.40), with the only difference that now the function $\varphi_i^{(3)}$ corresponds to the displacements of the right end.

Therefore, we write down only the coupling conditions for the segments:

$$\begin{aligned}
&\varphi_i^{(1)}(L_0) = \varphi_i^{(2)}(L_0) && \varphi_i^{(2)}(L_0+L_p) = \varphi_i^{(3)}(L_0+L_p)\\
&\varphi_i^{(1)'}(L_0) = \varphi_i^{(2)'}(L_0) && \varphi_i^{(2)'}(L_0+L_p) = \varphi_i^{(3)'}(L_0+L_p)\\
&\varphi_i^{(1)''}(L_0) = \frac{\mathrm{EI}^{(2)}}{\mathrm{EI}^{(1)}}\varphi_i^{(2)''}(L_0) && \varphi_i^{(2)''}(L_0+L_p) = \frac{\mathrm{EI}^{(1)}}{\mathrm{EI}^{(2)}}\varphi_i^{(3)''}(L_0+L_p)\\
&\varphi_i^{(1)'''}(L_0) = \frac{\mathrm{EI}^{(2)}}{\mathrm{EI}^{(1)}}\varphi^{(2)'''}(L_0) && \varphi_i^{(2)'''}(L_0+L_p) = \frac{\mathrm{EI}^{(1)}}{\mathrm{EI}^{(2)}}\varphi_i^{(3)'''}(L_0+L_p)
\end{aligned} \tag{1.49}$$

where $\mathrm{EI}^{(1)}$ is the flexural stiffness of the segment without a piezoelectric element, and $\mathrm{EI}^{(2)}$ is covered by a piezoelectric element.

Satisfying the boundary conditions, we obtain a homogeneous system of 12 equations with 12 unknowns:

$$\Lambda = \begin{pmatrix} a_{1,1} & \cdots & a_{1,12} \\ \vdots & \ddots & \vdots \\ a_{12,1} & \cdots & a_{12,12} \end{pmatrix}. \tag{1.50}$$

The specific mass m, bending stiffness EI, and function $J_p(x_1)$ are found from Eqs. (1.42), (1.44), and (1.44), respectively. But the function $G(x_1)$ in these equations for the given construction will have the form:

$$G(x_1) = H(x_1 - L_0) - H(x_1 - L_0 - L_p). \tag{1.51}$$

The method of dividing the substrate into three segments is also suitable for the case shown in Fig. 1.11, when the piezoelement begins at the clamping and does not completely cover the substrate, and the proof mass is displaced relative to the end.

In this case, the piecewise-defined function will look like this:

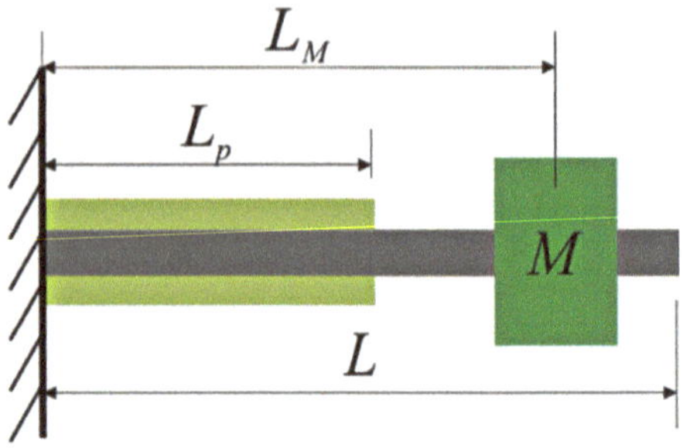

Fig. 1.11 Bimorph cantilever PEG with incomplete substrate coating with piezoelectric element and displaced proof mass

$$\phi_i(x_1) = \begin{cases} \phi_i^{(1)}(x_1), x_1 \le L_p \\ \phi_i^{(2)}(x_1), L_p < x_1 \le L_M \\ \phi_i^{(3)}(x_1), x_1 > L_M \end{cases} \tag{1.52}$$

The boundary conditions remain the same as in Eq. (1.49), only the coupling conditions of the segments differ:

$$\begin{array}{ll} \phi_i^{(1)}(L_0) = \phi_i^{(2)}(L_0) & \phi_i^{(2)}(L_M) = \phi_i^{(3)}(L_M) \\ \phi_i^{(1)\prime}(L_0) = \phi_i^{(2)\prime}(L_0) & \phi_i^{(2)\prime}(L_M) = \phi_i^{(3)\prime}(L_M) \\ \phi_i^{(1)\prime\prime}(L_0) = \dfrac{\mathrm{EI}^{(2)}}{\mathrm{EI}^{(1)}}\phi_i^{(2)\prime\prime}(L_0) & \begin{array}{l} \phi_i^{(2)\prime\prime}(L_M) = \varphi_i^{(3)\prime\prime}(L_M) \\ \phi_i^{(2)\prime\prime\prime}(L_M) = \phi_i^{(3)\prime\prime\prime}(L_M) - \alpha\beta^4\phi_i^{(2)}(L_M) \end{array} \\ \phi_i^{(1)\prime\prime\prime}(L_0) = \dfrac{\mathrm{EI}^{(2)}}{\mathrm{EI}^{(1)}}\phi_i^{(2)\prime\prime\prime}(L_0) & \alpha = \dfrac{M}{mL} \end{array} \tag{1.53}$$

where $\mathrm{EI}^{(1)}$ is the flexural stiffness of a segment covered by a piezoelectric element, and $\mathrm{EI}^{(2)}$ is without a piezoelectric element.

The function $G(x_1)$ for this construction will be equivalent to Eq. (1.43) because of the similarity of the construction.

Further, Figs. 1.12, 1.13, 1.14, and 1.15 show the dependencies of different characteristics of PEG on the length of the piezoelement for different thicknesses of piezoelements for the model shown in Fig. 1.11.

Analysis of Figs. 1.12, 1.13, 1.14, and 1.15 leads to similar conclusions to those obtained for configuration # 2. However, in this case, the presence of a proof mass of 3 g markedly lowered the values of the characteristics studied, except for the displacement of the end of the beam, which increased.

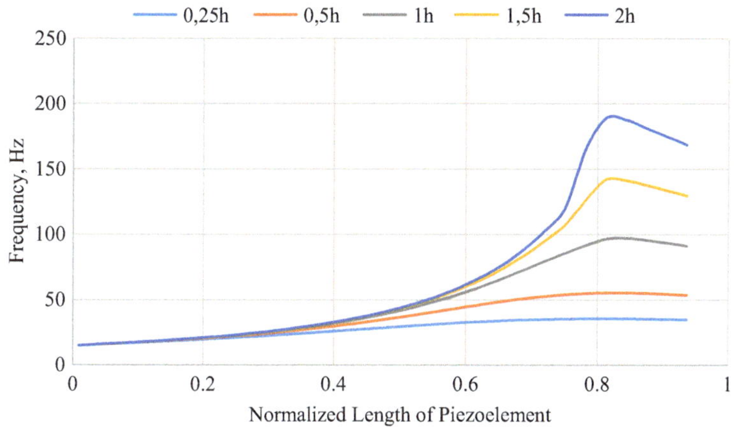

Fig. 1.12 First resonance frequency vs. normalized length of piezoelement for different thicknesses of the piezoelectric elements

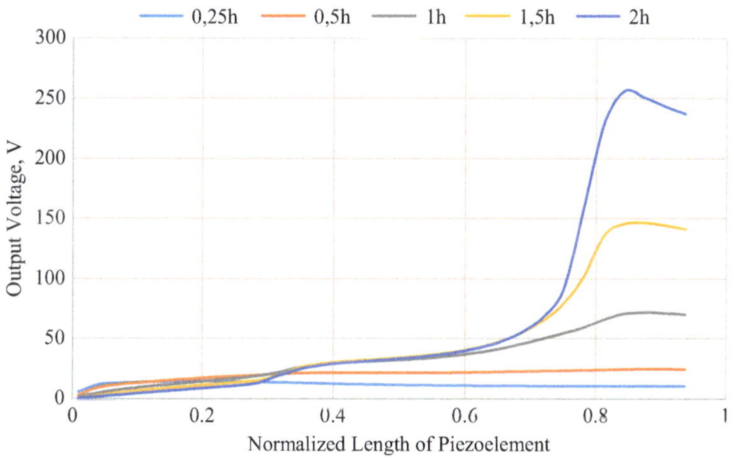

Fig. 1.13 Output voltage vs. normalized length of piezoelement for different thicknesses of the piezoelectric elements

1.2.2.4 Configuration # 4

It remains to consider the most general case (when the piezoelements have the same length relative to each other and are always symmetrically disposed), represented in Fig. 1.16.

The search for a solution for this design is associated with the need to divide the beam into four segments. The first segment is from the clamp to the beginning of the piezoelement. The second segment includes a part of the substrate covered with a

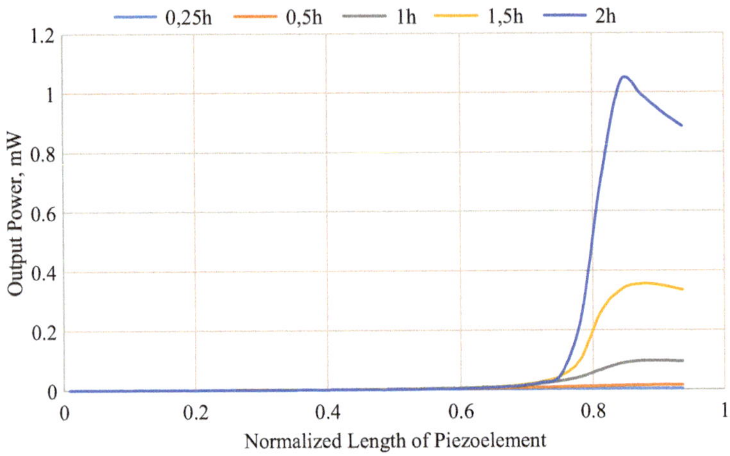

Fig. 1.14 Output power vs. normalized length of piezoelement fordifferent thicknesses of the piezoelectric elements

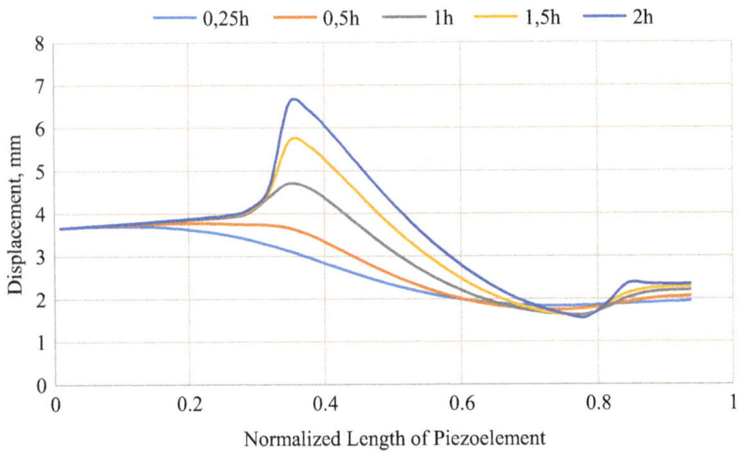

Fig. 1.15 Displacement of the beam's free end vs. normalized length of piezoelement for different thicknesses of the piezoelectric elements

Fig. 1.16 Bimorph cantilever PEG with displaced piezoelectric element relative to the clamp and displaced proof mass relative to the free end

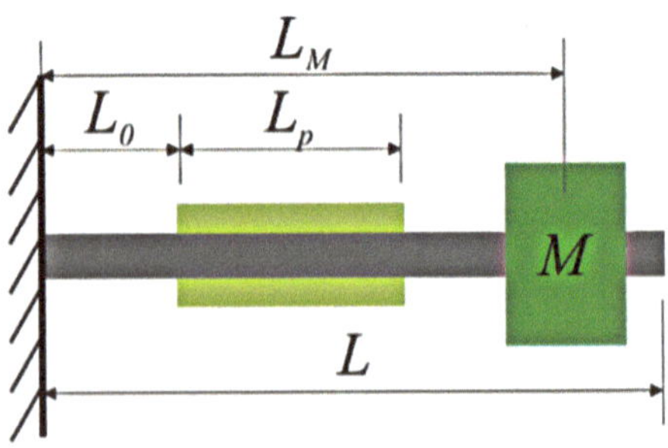

piezoelectric element. The third segment is the free part of the beam, following the piezoelement up to the attachment point of the proof mass. The fourth segment is after the proof mass up to the end of the beam. Taking into account the above division of the beam, the piecewise-defined function $\varphi_i(x_1)$ takes the following form:

$$\phi_i(x_1) = \begin{cases} \varphi_i^{(1)}(x_1), x_1 \leq L_0 \\ \varphi_i^{(2)}(x_1), L_0 < x_1 \leq L_p + L_0 \\ \varphi_i^{(3)}(x_1), L_p + L_0 < x_1 \leq L_M \\ \varphi_i^{(4)}(x_1), x_1 > L_M \end{cases} \tag{1.54}$$

where $\phi_i^{(1)}$, $\phi_i^{(2)}$, $\phi_i^{(3)}$, and $\phi_i^{(4)}$ correspond to the forms of oscillations of the first, second, third, and fourth segments, respectively. We write the solution in common form for each part of the beam:

$$\begin{aligned} \varphi_i^{(1)}(x_1) &= a_{1,i}\sin(\beta_i x_1) + a_{2,i}\cos(\beta_i x_1) + a_{3,i}\sinh(\beta_i x_1) + a_{4,i}\cosh(\beta_i x_1) \\ \varphi_i^{(2)}(x_1) &= a_{5,i}\sin(\beta_i x_1) + a_{6,i}\cos(\beta_i x_1) + a_{7,i}\sinh(\beta_i x_1) + a_{8,i}\cosh(\beta_i x_1) \\ \varphi_i^{(3)}(x_1) &= a_{9,i}\sin(\beta_i x_1) + a_{10,i}\cos(\beta_i x_1) + a_{11,i}\sinh(\beta_i x_1) + a_{12,i}\cosh(\beta_i x_1) \\ \varphi_i^{(4)}(x_1) &= a_{13,i}\sin(\beta_i x_1) + a_{14,i}\cos(\beta_i x_1) + a_{15,i}\sinh(\beta_i x_1) + a_{16,i}\cosh(\beta_i x_1) \end{aligned} \tag{1.55}$$

The boundary conditions are preserved, differing only by the function $\phi_i^{(4)}(x_1)$, which corresponds to the right end of the beam. The coupling conditions of the first and second segments, as well as the second and third are equivalent to Eq. (1.49). The coupling conditions of the third and fourth segments have the form:

$$\begin{aligned} &\phi_i^{(3)}(L_M) = \phi_i^{(4)}(L_M), \\ &\phi_i^{(3)\prime}(L_M) = \phi_i^{(4)\prime}(L_M), \\ &\phi_i^{(3)\prime\prime}(L_M) = \phi_i^{(4)\prime\prime}(L_M), \\ &\phi_i^{(3)\prime\prime\prime}(L_M) = \phi_i^{(4)\prime\prime\prime}(L_M) - \alpha\beta^4\phi_i^{(3)}(L_M), \\ &\alpha = \frac{M}{mL} \end{aligned} \tag{1.56}$$

Satisfying the boundary conditions, we obtain a homogeneous system of 16 equations with 16 unknowns:

$$\Lambda = \begin{pmatrix} a_{1,1} & \cdots & a_{1,16} \\ \vdots & \ddots & \vdots \\ a_{16,1} & \cdots & a_{16,16} \end{pmatrix} \tag{1.57}$$

The function $G(x_1)$ for this construction will be equivalent to Eq. (1.51) because of the similarity of the construction.

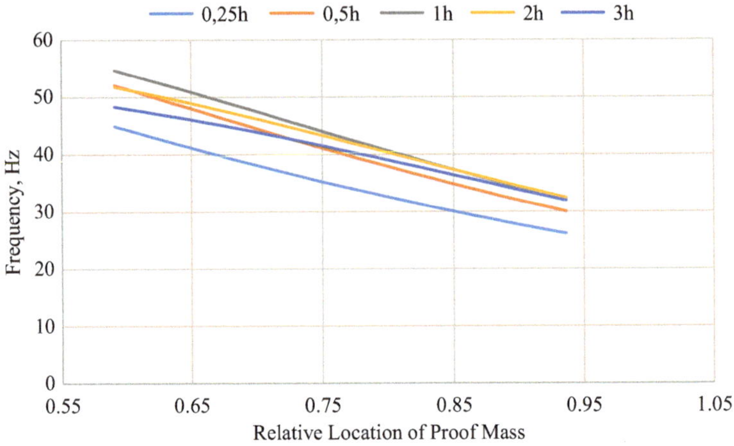

Fig. 1.17 First resonance frequency vs. relative location of proof mass for different thicknesses of the piezoelectric elements

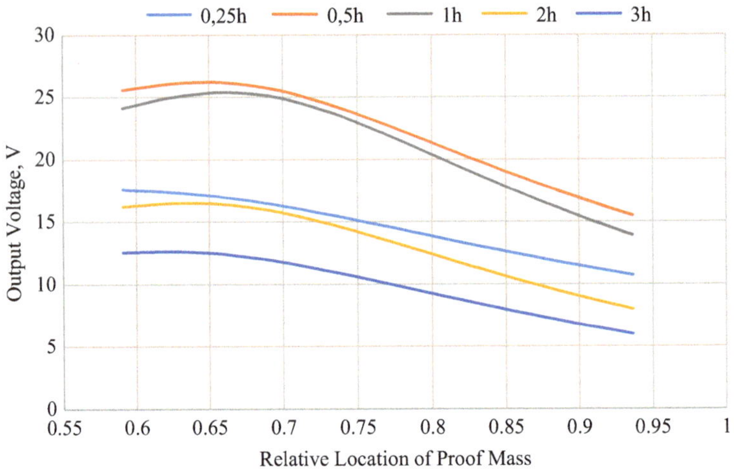

Fig. 1.18 Output voltage vs. relative location of proof mass for different thicknesses of the piezoelectric elements

Further, in Figs. 1.17, 1.18, 1.19 and 1.20, the dependencies of the main PEG performance characteristics are shown as a function of the position of the proof mass (3 g) for the model, shown in Fig. 1.16. For the convenience of analyzing the results, the position of the mass was normalized with respect to the length of the substrate. The change in the position of the proof mass occurred in the range between the point where the piezoelectric element ends and the free end of the substrate

From Figs. 1.17, 1.18, 1.19, and 1.20, the following conclusions can be drawn:

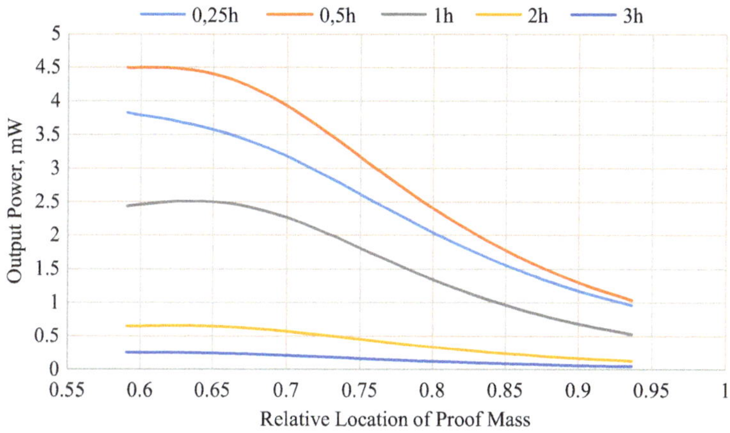

Fig. 1.19 Output power vs. relative location of proof mass for different thicknesses of the piezoelectric elements

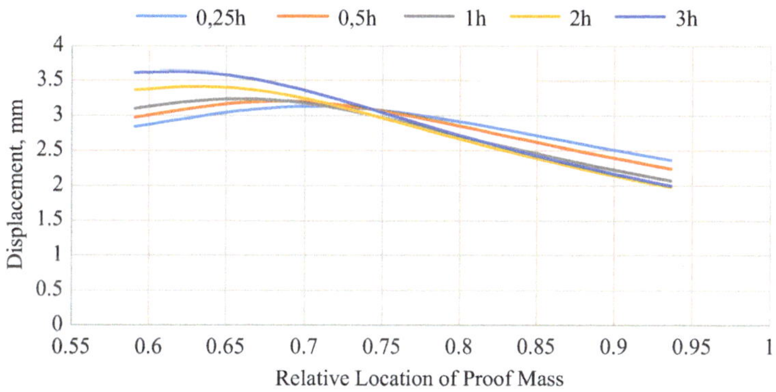

Fig. 1.20 Displacement of the beam's free end vs. relative location of proof mass for different thicknesses of the piezoelectric elements

(i) The further from the end of the piezoelement the proof mass is located, the lower the resonance frequency, the output voltage, and the output power;
(ii) The change in the thickness of the piezoelement has a nonlinear effect on all characteristics of the PEG;
(iii) The greatest output voltage and power are reached when the proof mass is located near the piezoelectric element, and the thickness of the piezoelectric element is half the thickness of the substrate;
(iv) The dependence of the displacement of the beam's end on the position of the proof mass is strongly affected by the thickness of the piezoelectric element; the thicker the piezoelectric element, the greater the displacement when the mass is located near the piezoelectric element and the smaller the displacement when the mass is located near the end of the beam.

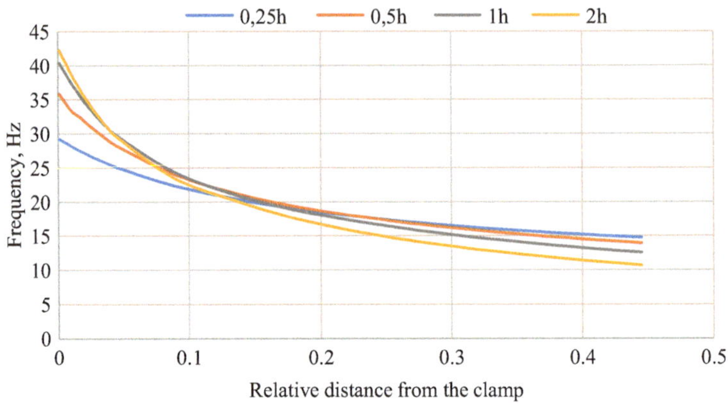

Fig. 1.21 First resonance frequency vs. relative distance from the clamp for different thicknesses of the piezoelectric elements

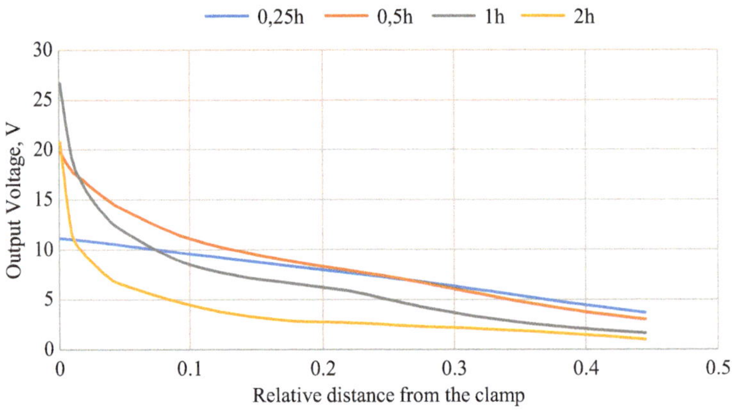

Fig. 1.22 Output voltage vs. relative distance from the clamp for different thicknesses of the piezoelectric elements

At the last stage, the influence of the position of a piezoelement of fixed length at various thicknesses on the different characteristics of PEG was investigated. As the main parameter determining the position of the piezoelectric element on the substrate, the distance from the clamp to the beginning of the piezoelectric element was taken. For the convenience of analysis, it was normalized with respect to the length of the substrate. The numerical results are shown in Figs. 1.21, 1.22, 1.23, and 1.24.

Analysis of the dependencies shown in Figs. 1.21, 1.22, 1.23, and 1.24, allows us to draw the following conclusions:

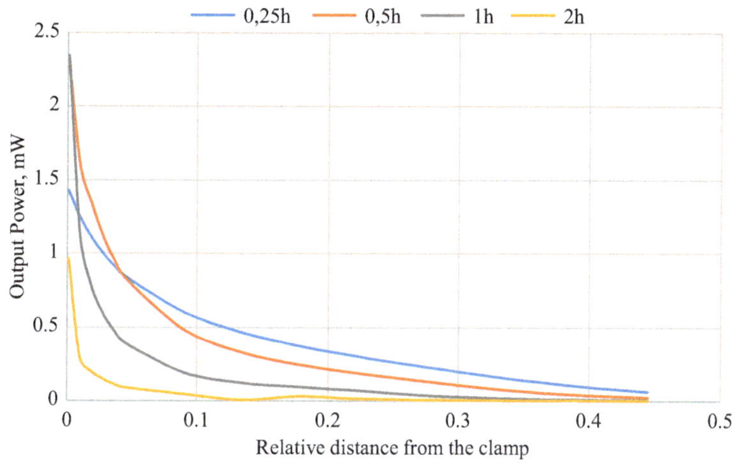

Fig. 1.23 Output power vs. relative distance from the clamp for different thicknesses of the piezoelectric elements

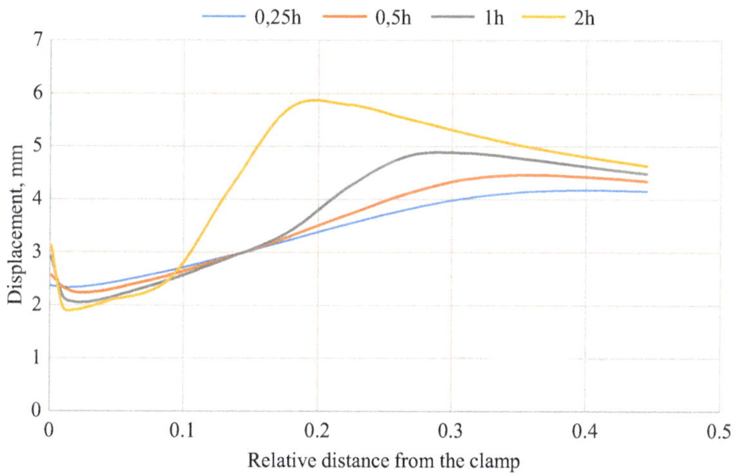

Fig. 1.24 Displacement of the beam's free end vs. relative distance from the clamp for different thicknesses of the piezoelectric elements

(i) The closer the piezoelectric elements are to the clamp, the higher the resonance frequency, the output voltage, and the output power;
(ii) The change in the thickness of the piezoelement has a nonlinear effect on all characteristics of the PEG;

Fig. 1.25 Stack-type PEG

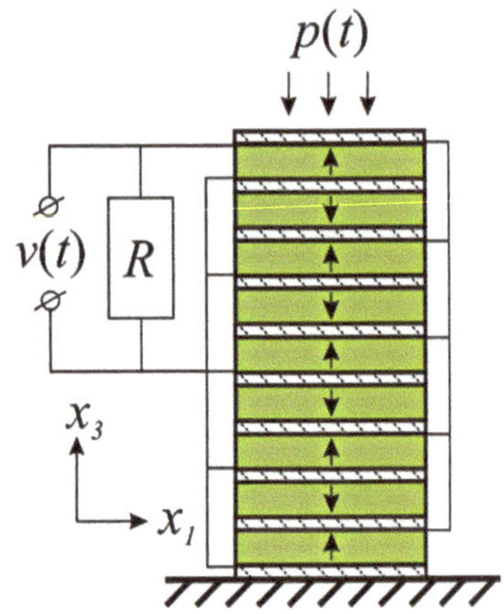

(iii) The greatest output voltage and power are reached when the piezoelements are located near the clamp, and the thickness of the piezoelectric element is equal to the thickness of the substrate;

(iv) The dependence of the displacement of the beam's end on the position of the piezoelectric element for various thicknesses has a complex shape; the thicker the piezoelectric element, the greater the displacement of the end of the beam, starting from a certain position of the piezoelectric element, before which a different picture is observed.

1.3 Mathematical Modelling of Stack-type PEGs

1.3.1 Statement of the Problem for Stack-type PEGs

The statement of the problem for the stack-type PEGs, shown in Fig. 1.25, is similar to the problem for the cantilever-type PEGs. Unlike the cantilever-type PEG, excitation of the stack-type PEG occurs through the application of a mechanical load $p(t)$ along the x_3 coordinate axis.

Therefore, Eqs. (1.9, 1.10, 1.11, 1.12, and 1.13) for the cantilever-type remain unchanged for stack-type PEGs.

To simplify the problem, it is convenient to consider it as forced longitudinal vibrations of the rod, excited by the external force $p(t)$ along x_3. To do this, we represent the displacement vector $\mathbf{u}$ in the following form:

$$\mathbf{u} = \{0, 0, w(x_3, t)\}^T \tag{1.58}$$

The transition to the consideration of the one-dimensional case also simplifies the defining relations (1.2):

$$\begin{aligned}\sigma_{33} &= c_{33}^{E}{*}\varepsilon_{33} - e_{33}{}^{*}E_3 \\ D_3 &= e_{33}{}^{*}\varepsilon_{33} + \ni_{33}^{S}{*}E_3\end{aligned} \tag{1.59}$$

where the material constants are expressed as follows:

$$c_{33}^{E*} = \frac{1}{s_{33}^{E}},\quad e_{33}{}^{*} = \frac{d_{33}}{s_{33}^{E}},\quad \ni_{33}^{S*} = \ni_{33}^{T} - \frac{d_{33}^{2}}{s_{33}^{E}} \tag{1.60}$$

In the PEG studied, the electrodes on each piezoelement are applied to long sides perpendicular to the x_3 axis, and therefore it makes sense to consider only the electric potential components along the x_3 axis.

It is assumed that the piezoelectric element is sufficiently thin, and there are no free charges inside it. Therefore, we introduce the assumption of a linear electric field distribution over the thickness of the piezoceramic element:

$$\phi_{,3} = \frac{v(t)}{h_p} \tag{1.61}$$

where $v(t)$ is the potential difference between the upper and lower electrode of the piezoelectric layer, h_p is the thickness of one piezoelectric layer.

Taking into account all the assumptions assumed above, Eq. (1.11) takes the following form:

$$\begin{aligned}\int_{t_1}^{t_2} dt \Bigg\{ \iiint_V \Bigg[&\left(-c_{33}^{E*}\frac{\partial w(x_3,t)}{\partial x_3} + e_{33}{}^{*}\frac{v(t)}{h_p}\right)\delta\left(\frac{\partial w(x_3,t)}{\partial x_3}\right) \\ &+ \left(\frac{e_{33}{}^{*}}{h_p}\frac{\partial w(x_3,t)}{\partial x_3} + \ni_{33}^{S*}\frac{v(t)}{h_p^2}\right)\delta v(t) \\ &- \rho\ddot{w}(x_3,t)\delta w(x_3,t)\Bigg] dV + \iint_S \left(p_3\delta w(x_3,t) + \frac{\sigma x_3}{h_p}\delta v(t)\right) dS \Bigg\} = 0\end{aligned} \tag{1.62}$$

To analyze this problem, it is also convenient to use the Kantorovich method, which was used earlier in the case of cantilever-type PEG. We represent the displacement of the rod, in the form of a series expansion:

$$w(x_3, t) = \sum_{i=1}^{N} \eta_i(t)\varphi_i(x_3) \quad (1.63)$$

where N is the number of modes considered, $\eta_i(t)$ are unknown generalized coordinates, and $\varphi_i(x_3)$ are known trial functions satisfying the boundary conditions.

Then, repeating the actions similar to those, used for the cantilever-type PEG, we obtain a system of differential equations. This system describes the electrical and mechanical response of a stack-type PEG, connected to a resistor to the excitation of an external mechanical force:

$$\begin{aligned} &\mathbf{M}\ddot{\boldsymbol{\eta}}(t) + \mathbf{D}\dot{\boldsymbol{\eta}}(t) + \mathbf{K}\boldsymbol{\eta}(t) - \boldsymbol{\Theta}v(t) = \mathbf{p} \\ &C_f\dot{v}(t) + \boldsymbol{\Theta}^{\mathrm{T}}\dot{\boldsymbol{\eta}}(t) + \frac{v(t)}{R} = 0 \end{aligned} \quad (1.64)$$

The resulting system is equivalent to the system shown in Eq. (1.21), but the coefficients for the variables are different:

$$\begin{aligned} &C_{\Pi Э} = N\frac{b_p l_p}{h_p} Э_{33}^{S*} && p_i = -p_0\varphi_i(x_3) \\ &M_{ij} = \int_0^H m\varphi_i(x_3)\varphi_j(x_3)dx_3 && \theta_i = \int_0^H J_p\varphi_i'(x_3)dx_3 \\ &K_{ij} = \int_0^H Y\varphi_i'(x_3)\varphi_j'(x_3)dx_3 && Y = \iint_{S_p} c_{33}^{E*}dS \\ & && J_p = \iint_{S_p} \frac{e_{33}{}^*}{h_p}dS \end{aligned} \quad (1.65)$$

Now we solve the problem of finding the trial functions, which satisfy the boundary conditions. Since the system in Eq. (1.64) is equivalent to the system in Eq. (1.21), the solution for the harmonic case will be the same as in Eq. (1.23).

In real operating conditions, exciting force $p(t)$ has a complex shape. Therefore, we consider this case. One of the approaches that allow us to define functions by an arbitrary form is the approximation by means of Fourier series. To do this, we represent $p(t)$ with a set of discrete values, and then perform an interpolation by means of Fourier series:

$$p(t) \cong m_0 + \sum_{k=1}^{N}\left[m_k\cos\left(k\frac{2\pi t}{T}\right) + n_k\sin\left(k\frac{2\pi t}{T}\right)\right], \quad (1.66)$$

where m_0 is the mean value, T is the time range of loading, and n_k, m_k are the Fourier coefficients.

$$m_0 = \frac{1}{T}\int_0^T p(t)dt, \quad m_k = \frac{2}{T}\int_0^T p(t)\cos\left(k\frac{2\pi t}{T}\right)dt,$$
$$n_k = \frac{2}{T}\int_0^T p(t)\sin\left(k\frac{2\pi t}{T}\right)dt \tag{1.67}$$

Further, we substitute the resulting representation (1.66) into Eq. (1.64). We get a system of differential equations, whose solution in explicit form is rather problematic. Therefore, we will use the Runge-Kutta numerical method to solve it.

1.3.2 Numerical Experiment

The Kantorovich method we used recommends finding trial functions that satisfy the boundary conditions (in our case, mechanical). To do this, we solve the eigenvalue problem for the longitudinal vibrations of the rod shown in Fig. 1.25.

Let us write out the solution of the equation of longitudinal oscillations in general form:

$$\phi_i(x_3) = a_{1,i}\sin(\beta_i x_3) + a_{2,i}\cos(\beta_i x_3) \tag{1.68}$$

Satisfying the boundary conditions:

$$\varphi_i(0) = 0\phi_i'(H) = 0 \tag{1.69}$$

we can find the eigenvalues β_i and the coefficients a_i.

After this, we obtain a homogeneous system of four equations with four unknowns. We write it in the matrix form:

$$\Lambda = \begin{pmatrix} a_{1,1} & \cdots & a_{1,4} \\ \vdots & \ddots & \vdots \\ a_{4,1} & \cdots & a_{4,4} \end{pmatrix} - 0 \tag{1.70}$$

This system has non-zero solutions, when its determinant is zero. The determinant of the system is the simplest trigonometric equation:

$$\cos\beta_i = 0 \tag{1.71}$$

Knowing β_i, we can find the factors a_i for the required number of vibration modes N.

Since the resonance frequency of this PEG type is high, it is usually used at frequencies well-below the first resonance frequency. In addition, its ability to work at high mechanical loads makes it possible to obtain sufficiently high values of the output voltage even at low frequencies.

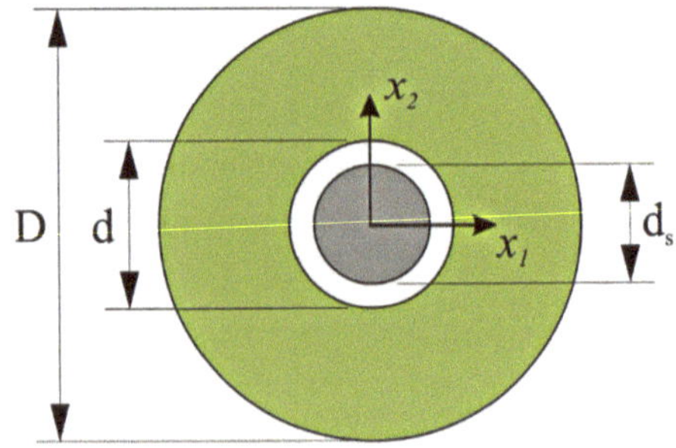

Fig. 1.26 The cross-section of the stack-type PEG

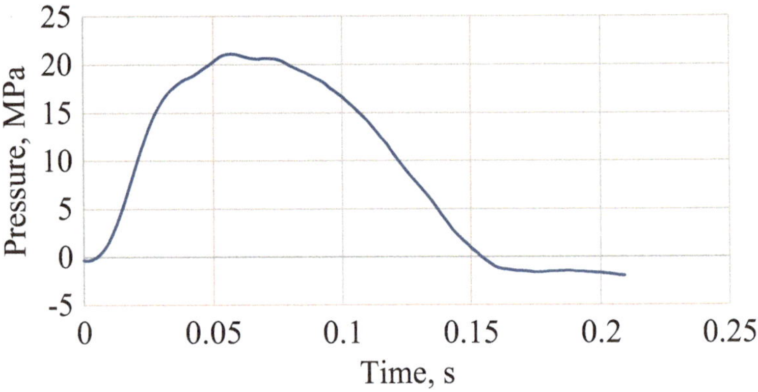

Fig. 1.27 Form of loading force

We will consider the stack-type PEG, shown in Fig. 1.25. However, it will consist of annular piezoceramic elements made of PZT-19 piezoceramics. It is also assumed that the set of elements is tightened between a coupling bolt. In this regard, the cross-section will have the form shown in Fig. 1.26. To account for the effect of the metal core (coupling bolt) in the PEG cross-section, we represent the stiffness of the section Y as follows:

$$Y = \iint\limits_{S_p} c_{33}^{E} * dS + \iint\limits_{S_s} C_s dS, \tag{1.72}$$

where S_s is the cross-section area of the bolt, C_s is the modulus of elasticity of the steel.

This PEG will be excited by the load, the shape of which is presented in Fig. 1.27. The main geometric and physical properties of the generator model under study are given in Table 1.2. The coefficients of the modal damping are the same as in the cantilever-type PEG.

Within the framework of the developed model, a number of numerical experiments were conducted. Dependencies of electrical characteristics of stack-type PEGs on various parameters were investigated. The results are shown in Figs. 1.28, 1.29, 1.30, and 1.31.

Table 1.2 PEG parameters

	Coupling bolt	Piezoelement
Geometrical dimensions ($D \times d \times h$)	6 mm	18 × 8 × 1 mm^3
Density (ρ)	7,800 kg/m^3	7,500 kg/m^3
The Young's modulus and Poisson's ratio (E, ν)	210 GPa and 0.3	–
Elastic compliance $\left(s_{33}^E\right)$	–	17.5 × 10^{-12} Pa
Relative permittivity $\left(\varepsilon_{33}^S/\varepsilon_0\right)$	–	1,500
Piezoelectric module (d_{33})	–	−307 pC/N

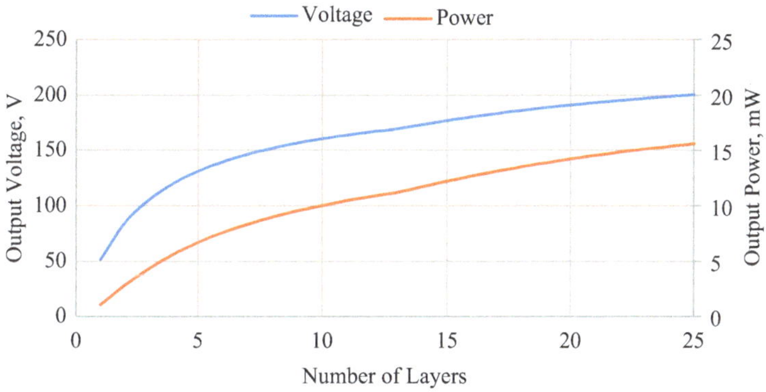

Fig. 1.28 Dependence of output voltage and output power of stack-type PEG on the number of layers

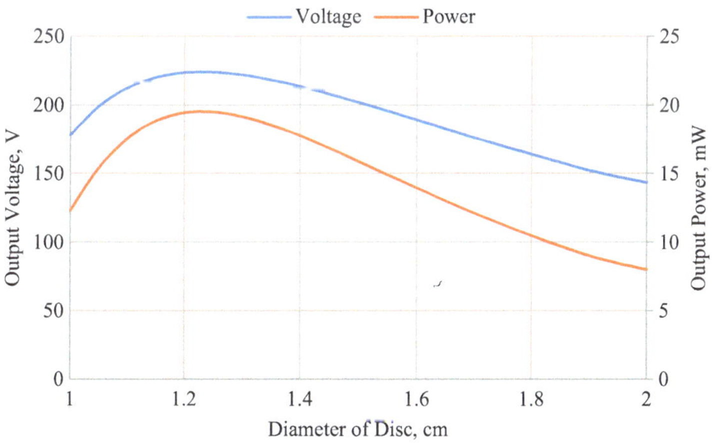

Fig. 1.29 Dependence of the output voltage and output power of the stack-type PEG on the diameter of the piezoelectric element

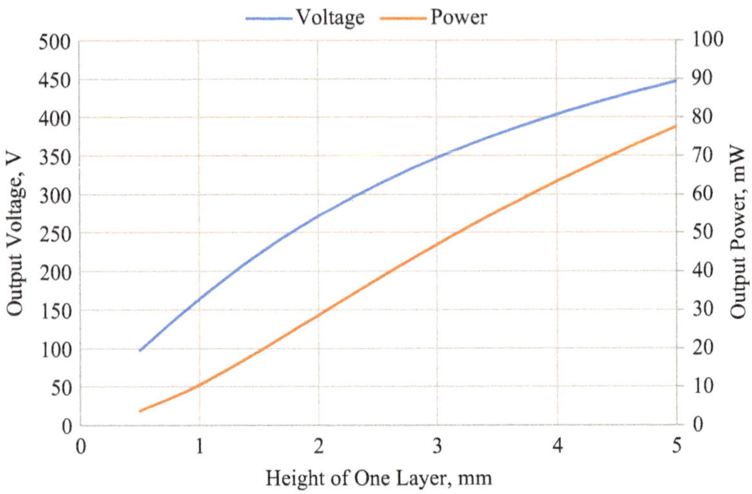

Fig. 1.30 Dependence of output voltage and output power of stack-type PEG on the height of each layer

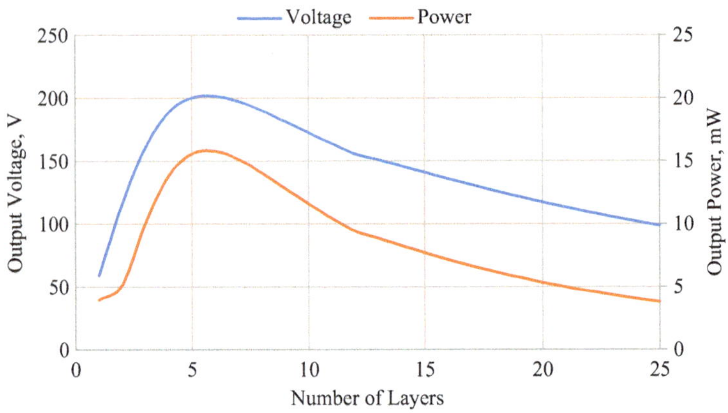

Fig. 1.31 Dependence of output voltage and output power of stack-type PEG on the number of layers, provided that the height of the entire packet remains unchanged

Analysis of Figs. 1.28, 1.29, 1.30, and 1.31 allows us to draw the following conclusions:

(i) with an increase in the number of layers, the output voltage and power increase;
(ii) with an increase in the external diameter of the disk, the output voltage and power increase to a certain value, after which they decrease;
(iii) with an increase in the height of the layers, the output voltage and power increase;
(iv) at a fixed total height of the entire piezo-package (the height of each layer depends on the total number of layers), there is a value for the number of layers, at which the output characteristics will be maximum.

Chapter 2
Experimental Modeling of Piezoelectric Generators

2.1 Cantilever-Type Generators

This section describes an experiment to determine the performance of a laboratory sample of a cantilever-type piezoelectric generator (PEG). The test set-up, experimental procedure, and experimental data are presented. In addition, the results of the mathematical model described in the previous chapter are compared with the data obtained during the experiment.

2.1.1 *Description of Test Set-up and Samples*

The output voltages of PEGs at different excitation frequencies are determined by a specially designed test bench [14]. Mechanical excitation is performed by the harmonic movement of the PEG clamp. The output voltage is measured at different values of load resistance at the resonance frequency.

The measuring set-up (see Fig. 2.1) consists of the electromagnetic shaker VEB Robotron 11077 (Fig. 2.1—see point 6) on the working table (Fig. 2.1—see point 5) 5, on which there are installed the cantilever PEG, the optical sensor of mechanical movements (Fig. 2.1—see point 7), the optical sensor controller (Fig. 2.1—see point 8) , the acceleration sensor (Fig. 2.1—see point 9) on the clamp surface (Fig. 2.1—see point 4), the acceleration sensor controller (Fig. 2.1—see point 10), the external module of ADC/DAC E14-440D of L-Card (Fig. 2.1—see point 11), the power amplifier LV-102 up to 50 W (Fig. 2.1—see point 12), the functional signal generator AFG 3022B Tektronix (Fig. 2.1—see point 13), computer (Fig. 2.1—see point 14), and load resistance (Fig. 2.1—see point *R*).

The electric response to mechanical loading is investigated, the output voltage of which is applied to the ADC input on the load resistor *R*. The optical sensor measures

S. N. Shevtsov et al., *Piezoelectric Actuators and Generators for Energy Harvesting*, Innovation and Discovery in Russian Science and Engineering,
https://doi.org/10.1007/978-3-319-75629-5_2

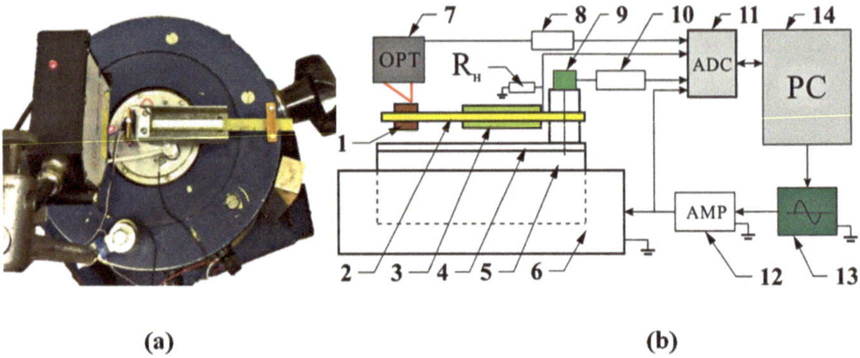

Fig. 2.1 General view of measuring set-up: (**a**) photograph of the set-up; (**b**) a block diagram: 1—proof mass, 2—PEG substrate, 3—bimorph piezoelectric element, 4—piezogenerator base, 5—shaker's working table, 6—shaker, 7—optical linear sensor, 8—optical sensor controller, 9—acceleration sensor, 10—acceleration sensor matching device, 11—external ADC/DAC module, 12—power amplifier, 13—signal generator, 14—computer, *R*– electrical resistance

the mechanical movements of the monitored surface without coming into contact with the surface. The optical sensor's matching device is its controller. The acceleration of the desktop is measured by the ADXL-103 acceleration sensor.

The cantilever-type PEG that was used as an experimental sample was made on the base of a glass-textolite substrate (see Fig. 2.1). The dimensions of the substrate (length × thickness × width) are $108 \times 10 \times 1$ mm^3. Piezoelements with dimensions $56 \times 6 \times 0.5$ mm^3, made using ceramics of PCR-7 M (with a piezoelectric constant $d_{31} = 350$ pC/N) and connected by a bimorph scheme, were glued with both sides of the substrate. The value of the proof mass was 3 g. Its initial position was at the free end of the beam.

2.1.2 *Experiment*

The purpose of the experiment was to study the influence of the position of the proof mass on the output characteristics of PEG at various discrete values of electrical resistance. The position of the proof mass relative to the clamp varied from 65 mm to 103 mm. At the first step, the position of the proof mass was fixed. Then, by scanning the frequencies of the exciting signal, the resonance frequency of bending vibrations was found. At the next step, the output electric voltage was measured at different values of electrical resistance. At the final step, the position of the proof mass was changed and the whole process began again at the first step. The results of the experiment are present in Figs. 2.2, 2.3 and 2.4.

Figure 2.2 shows the dependence of the first resonance frequency on the position of proof mass relative to the clamp. As the distance between the proof mass and the

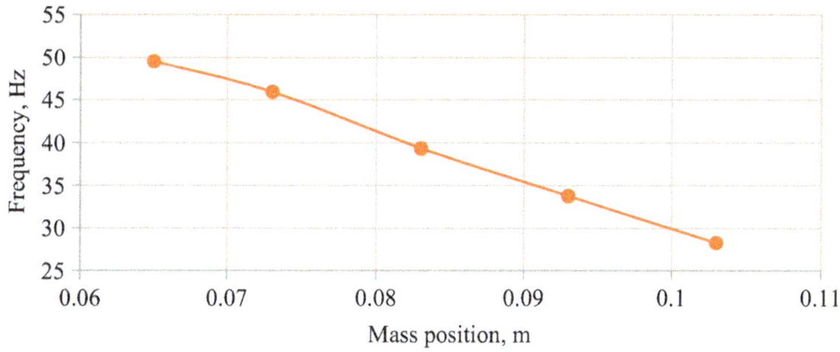

Fig. 2.2 Dependence of first resonance frequency on the position of proof mass

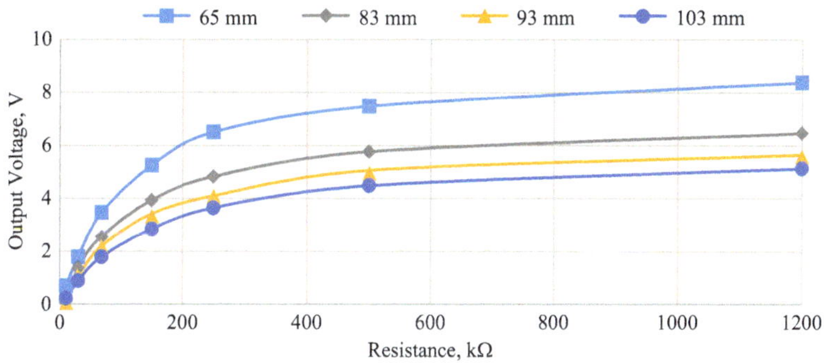

Fig. 2.3 Dependence of output voltage on electrical resistance at different positions of proof mass

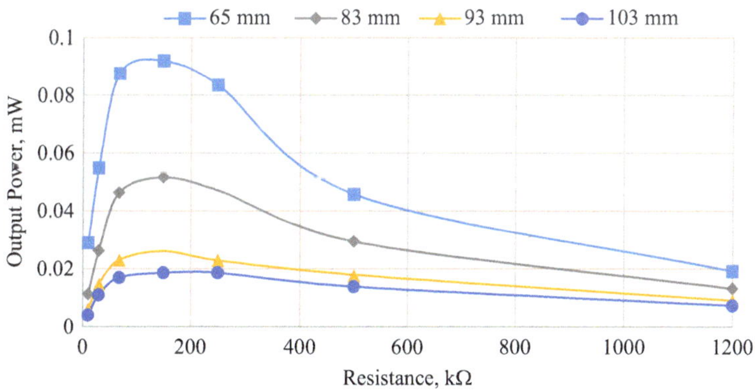

Fig. 2.4 Dependence of output power on electrical resistance at different positions of proof mass

clamp increases, the first resonance frequency of the flexural vibrations of the beam decreases.

Figure demonstrates the dependence of output voltage on the electrical resistance at different positions of the proof mass relative to the clamp. We can draw the following conclusions. Firstly, as the electrical resistance increases, the output voltage increases. Secondly, as the distance from the center of the proof mass to the point of clamping increases, the output electric voltage decreases for a fixed value of the electrical resistance.

From Fig. 2.4, where the dependence of output power on electrical resistance at different positions of proof mass relative to the clamp is illustrated, it follows that for each position of the proof mass, there is a maximum value of the output power. In addition, for a fixed value of electrical resistance, the output power decreases with increasing distance between the clamp and the proof mass, as in the case of the output voltage.

2.1.3 Theory and Experiment

Using the data obtained in the experiment, we compare these with the results of theoretical calculations based on the model described in the previous chapter. The combined plots of the dependencies, obtained during the experiment and theoretical calculations, are presented in Figs. 2.5, 2.6, and 2.7.

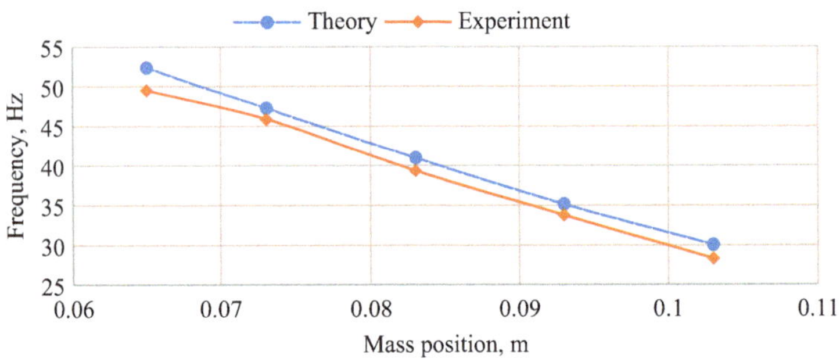

Fig. 2.5 Dependence of first resonant frequency on the position of proof mass: t—theory (dashed line) and e—experiment (solid line)

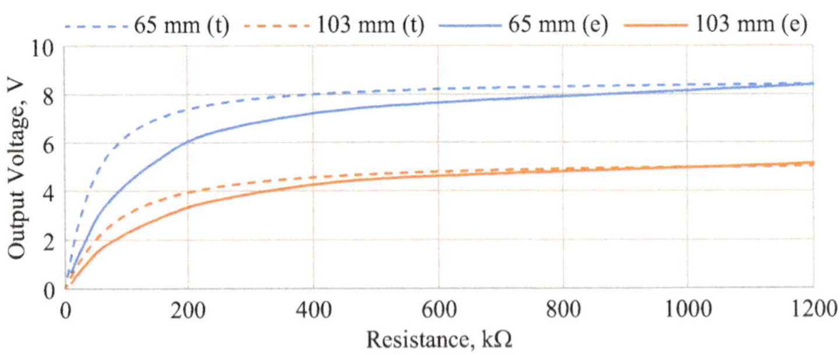

Fig. 2.6 Dependence of output voltage on electrical resistance at different positions of proof mass: t—theory (dashed line) and e—experiment (solid line)

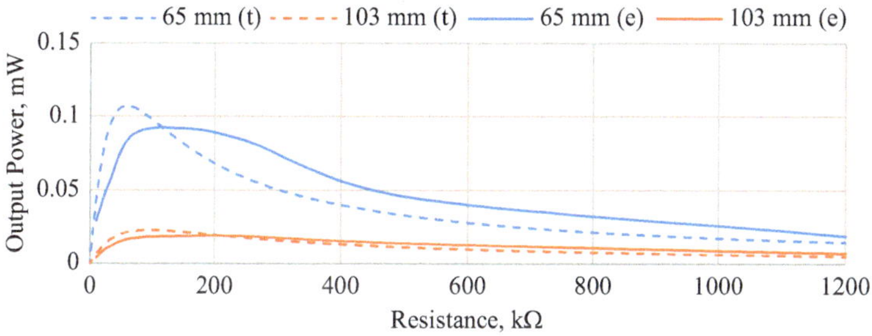

Fig. 2.7 Dependence of output power on electrical resistance at different positions of proof mass: t—theory (dashed line), and e—experiment (solid line)

From Fig. 2.5, where a comparison of theoretical and experimental first bending frequencies is presented for different positions of the proof mass, it follows that the model demonstrates good convergence with the experiment. The error does not exceed 5%.

Dependencies in Fig. 2.6, showing the difference between the experimental and calculated values of the output voltage, show that the calculated data are sufficiently close to the experimental data. The arithmetic mean error was 18%. This variation between the experiment and theory is due to the difference between the calculated and actual capacity of piezoelements, which is determined by many production factors.

2.2 Stack-Type Generators

This section describes experiments to determine the output performance of stack-type PEGs. Three types of loading are considered: harmonic, impulse, and quasi-static. For each loading type, the experimental set-up, experimental procedure, and the results are described. A comparison is made with theoretical calculations.

2.2.1 Harmonic Loading

2.2.1.1 Description of Test Set-up and Samples

The structural diagram of the set-up is shown in Fig. 2.8. The mechanical loading of the test specimen is carried out by the loading module of the set-up, consisting of an electric motor with a gearbox with an eccentric driver and a crank mechanism. To regulate the engine speed, the frequency converter VFD004L21A is used, which can be controlled by a computer. The stand includes a loading module with a geared motor, a strain gauge dynamometer, a strain gauge amplifier, a voltage converter

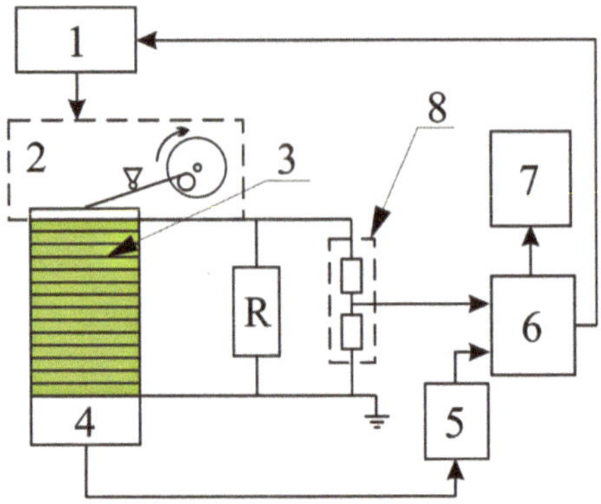

Fig. 2.8 Structural diagram of the test bench: 1—frequency converter VFD004L21A, specifying the frequency of the mechanical load; 2—loading module; 3—test sample of piezoelectric generator; 4 – tensometric dynamometer; 5 –strain gauge amplifier; 6—ADC/DAC; 7—PC;8—voltage divider

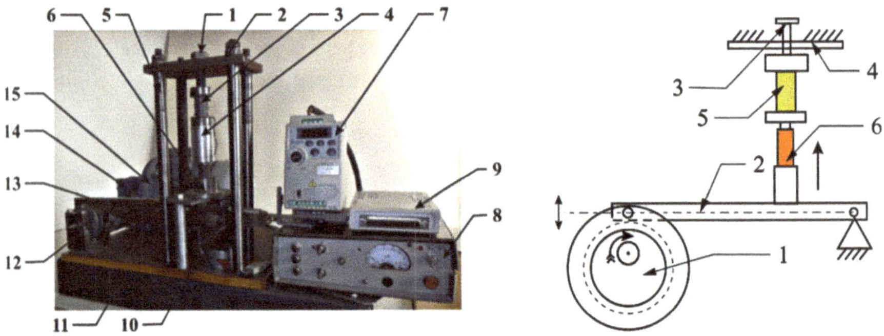

(a) **(b)**

Fig. 2.9 Test bench: (**a**) Photograph of the laboratory test bench for determining the characteristics of the stack-type piezoelectric generator : 1—screw, intended for setting the height of the generator and its initial preload, 2—fixed traverse, 3—piezogenerator, 4—tensometric dynamometer, 5—power columns, 6—guide cylinder with movable traverse, 7—frequency converter, 8—strain gauge, 9—ADC/DAC converter, 10—support bracket, 11—base, 12—eccentric disk with the connecting rod (13), 14—reducer, 15—electric motor; (**b**) kinematic scheme of the set-up: 1—electric motor, reducer and eccentric disk, 2—crank mechanism, 3—clamping screw, 4—traverse, 5—piezogenerator, 6—tensometric dynamometer

measuring ADC/DAC, an electrical resistance *R*, a voltage divider, a frequency converter specifying the frequency of a mechanical harmonic load, and a computer. The photograph of the test bench and its components is shown in Fig. 2.9a.

The laboratory set-up provides mechanical low-frequency loading of the PEG during harmonic excitation of the investigated PEG sample [57]. Excitation can be carried out at the modes of program and manual control of the amplitude and frequency components of the force, as well as the registration of the input and output parameters of the impact and response.

The work of the loading module of the set-up is explained by the kinematic scheme shown in Fig. 2.9b.

To measure electrical signals from a tensometric dynamometer and a piezogenerator, and also to process analog and digital information, the converter included a voltage measuring ADC/DAC E14-140 and Power Graph software.

The output voltage of PEGs at various loading frequencies (0.3–4 Hz) is determined on a test bench. The mechanical loading of the PEG is achieved at a compressive harmonic force with an amplitude from 0.1 to 5 kN. For each cycle of mechanical force, determined by means of a dynamometer, the time display of the magnitude of the mechanical force and the output voltage for various discrete values of the load resistance of R from 1 kΩ to 50 MΩ is recorded.

2.2.1.2 Experiment

Multilayer PEGs of prismatic geometry with a cross-sectional dimension of 24 × 16 mm^2 and various heights (36 mm and 21 mm) were studied. Multilayer PEGs are made of piezoelectric ceramics, based on piezoceramics PZT-19 M with 0.5 mm thickness, electrodes applied to the outer surfaces of the PEG and connected in parallel. Sintered using traditional ceramic technology, piezoelements were polarized in height $d_{33} = 360$ pC/N. One of the tested PEG samples is shown in Fig. 2.10.

Based on the experimental data for the PEG sample of 24 × 16 × 36 mm^3, the plots of the dependencies of the output voltage on the loading frequency and different values of the electrical load were plotted for the maximum value of the mechanical load (3.4 kN) (see Fig. 2.11).

As can be seen in Fig. 2.12, the maximum value of the output voltage (22.5 V) is achieved at a loading frequency of 4 Hz and a load resistance of 1.2 MΩ.

Fig. 2.10 Experimental sample of a prism-type PEG

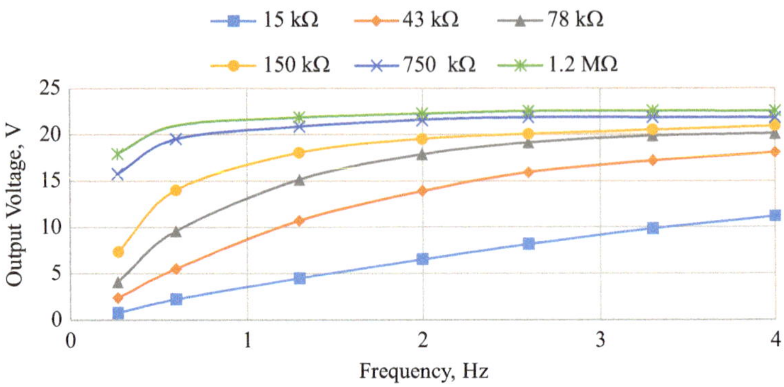

Fig. 2.11 Dependence of output voltage of the generator with the sizes of 24 × 16 × 36 mm^3, on the frequency of harmonic mechanical loading at different values of electrical resistance

The dependence of the electric voltage is almost linear in the load resistance of no more than 15 kΩ in the frequency range up to 4 Hz.

Further, based on the results of the measured peak values of the output voltage and the corresponding values of the electric resistance *R*, the maximum output power of the PEG was 3 mW.

Similar measurements were made for a PEG with dimensions of 24 × 16 × 21 mm^3 at frequencies of 2.6, 3.3 and 4.0 Hz and a load resistance from 6 to 750 kΩ.

From the analysis of the dependencies in Fig. 2.12, it can be concluded that as the electrical resistance increases to 78 kΩ, the output voltage increases. However, after this electrical resistance, only a slight increase is observed. The maximum value of output voltage obtained during the experiment was 22 V at an electrical resistance of 750 kΩ.

2.2.1.3 Comparison of Theory and Experiment

The data obtained for PEGs with heights of 21 mm and 36 mm were compared with each other, as well as with the results of theoretical calculations, based on the model, described in the previous chapter for the case of harmonic loading. The graphs of these dependencies for a frequency of 4 Hz are shown in Fig. 2.13.

From the comparative analysis of the experimental and theoretical results shown in Fig. 2.13, it can be concluded that for a PEG of 36 mm height, the difference between the experimental and theoretical data does not exceed 6%. For PEGs of 21 mm height, the difference was 16%, which is due to the fact that a modified composition of ceramics was used for its production, for which there is still no complete set of material constants.

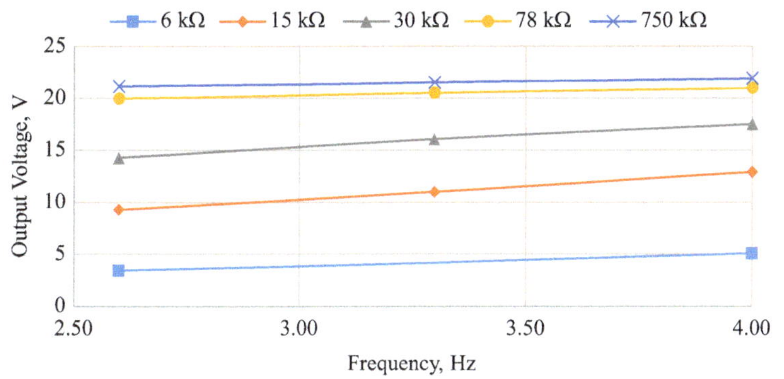

Fig. 2.12 Dependence of output voltage of the generator with the sizes of 24 × 16 × 21 mm^3 on the frequency of harmonic mechanical loading at different values of electrical resistance

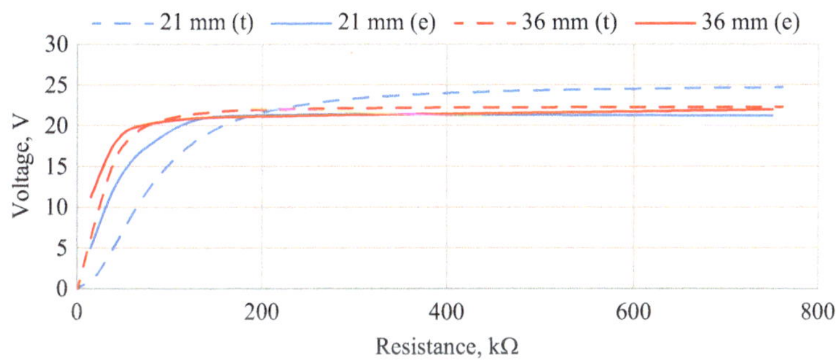

Fig. 2.13 Comparison of calculated (theoretical) and experimental data for PEGs of 21 and 36 mm in height with harmonic loading at a frequency of 4 Hz: t—theory (dashed line), e—experiment (solid line)

2.2.2 *Pulsed Loading*

The next type of loading of stack-type PEGs is pulsed loading. This loading type, like the previous one, is often found in various designs and techniques. In particular, such loading is characteristic for oscillations of rails in railway transport or a highway roadway. Due to the lack of standard experimental devices designed for stack-type PEG studies with low-frequency impulse loading, it was necessary to create non-standardized measuring instruments, namely a test bench for pulsed PEG loading [57].

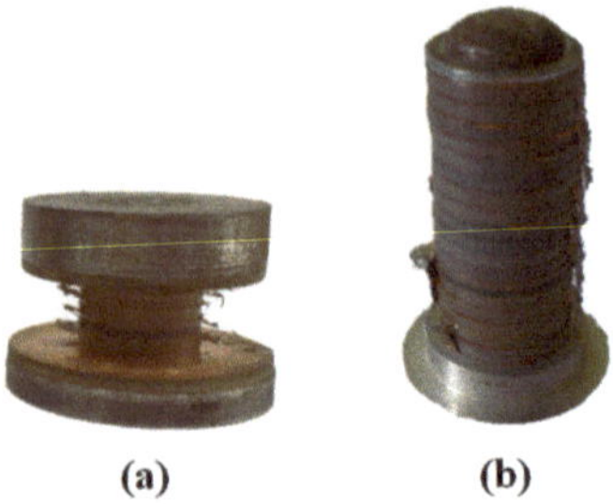

Fig. 2.14 Experimental samples of PEGs with ring sections: (**a**) PEG of 21-mm height, (**b**) PEG of 36-mm height

2.2.2.1 Description of the Test Set-up and Samples

Two PEGs of the ring-type were studied. The first PEG consisted of 11 disk piezoelements connected in parallel , each with 1-mm thickness. The second PEG consisted of 16 elements with 2-mm thickness. The inner and outer diameters of the rings in both PEGs were 18 mm and 8 mm, respectively. At the same time, each of the piezoelements was polarized in thickness ($d_{33} = 360$ pC/N), as a PEG material, piezoceramic PZT-19, was used. The electrical capacity of the first PEG was 20.22 nF and the second PEG had an electrical capacity of 21.3 nF. Photographs of the tested samples are shown in Fig. 2.14.

The experiment was carried out using the modernized laboratory set-up. This set-up provides mechanical low-frequency pulsed loading of PEGs at the modes of manual and manual control of the amplitude and frequency components of the force with registration of the input and output parameters of the impact and response. The structure and working principle of this set-up are similar to the set-up for creating harmonic loading (see Fig. 2.8). The principal difference lies in the design features of the stand.

Figure 2.15a shows a photograph of the loading module of the laboratory set-up. The main difference between this set-up and the set-up with harmonic loading is the features of the loading module. This is explained by the kinematic scheme shown in Fig. 2.15b.

The procedure for carrying out the experiment is similar to the method for harmonic loading described in the previous section. Mechanical cyclic loading of the PEG was carried out by a compressive pulsed force with an amplitude from 1 to 4 kN. For each cycle of the compressive force (determined with a dynamometer) the time dependencies of the magnitude of the compressive force and the output voltage were recorded for various discrete values of the load resistance R from 10 kΩ to 22.8 MΩ.

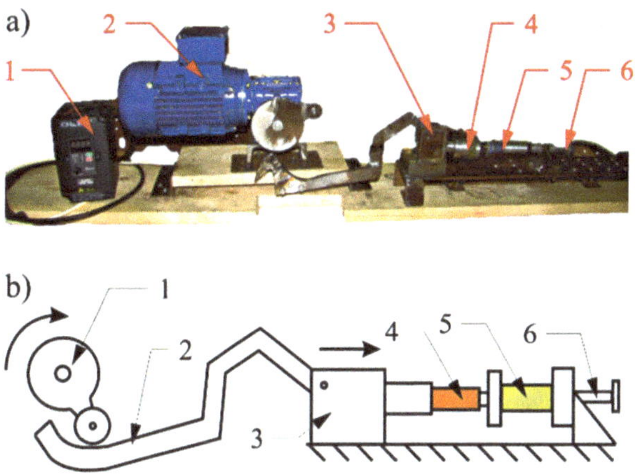

Fig. 2.15 Set-up for pulsed loading: (**a**) Photograph of the laboratory test bench for determining the characteristics of the piezoelectric generator of the stack-type: 1—frequency converter; 2—engine with gearbox, eccentric disk, and connecting rod; 3—lever multiplier of the compressive force change with a conversion factor of 50; 4—PEG test sample; 5—tensometric dynamometer; 6—PEG mounting bracket; (**b**) kinematic scheme of the loading module: 1—electric motor, gearbox, and eccentric disk; 2—lever; 3—conversion mechanism; 4—piezogenerator; 5—tensometric dynamometer; 6—clamping screw

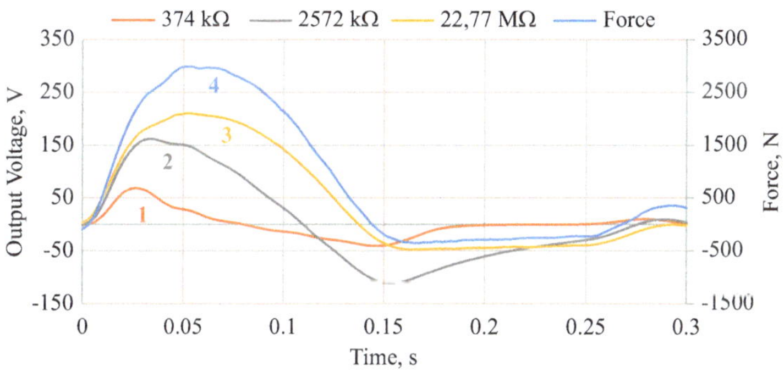

Fig. 2.16 Forms of compression force and output voltage for PEG sample with a height of 11 mm: curves 1, 2, and 3—voltage for load resistance 0.374; 2.572, and 22.77 MΩ, respectively, 4—force

2.2.2.2 Experiment

In accordance with the described method for measuring the output characteristics of a stack-type PEG, the dependence of output voltage on time for a PEG of height 11 mm was tested. The output voltage was obtained under axial impulse loading with an amplitude of 17.2 MPa (3 kN) and different values of the electric resistance R. Figure 2.16 shows the time dependencies of the output voltage of the

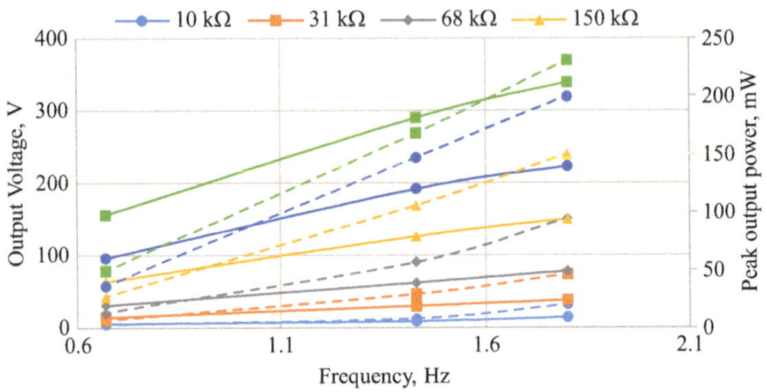

Fig. 2.17 Dependence of the output voltage (solid line) and the output electric power (dashed line) for different values of the electrical resistance from the loading frequency for ring-type PEGs with a height of 36 mm

dynamometer, corresponding to the compressive force 3 kN (curve 4) and piezoelectric responses at $R = 0.374$, 2.572, and 22.77 MΩ (curves 1, 2 and 3) for multilayer PEG.

Figure 2.16 demonstrates the effect of the load resistance on the correspondence of the piezoelectric response forms to the shape of the force as R increases. As can be seen from the plots in Fig. 2.16 for $R = 22.77$ MΩ, the amplitudes of the compressive force and the output voltage simultaneously reach a maximum at $t = 0.05$ s.

For a PEG sample with a height of 36 mm at a force of 3 kN and different values of the electrical resistance R, the dependencies of the output voltage on the frequency of mechanical loading were tested.

As follows from the plots of these dependencies in Fig. 2.17 (solid lines), the output electrical voltage of the PEG increases monotonically as the frequency of its loading increases and shows an almost linear trend at values load resistance from 68 to 500 kΩ. The maximum output voltage was 327 V at an electrical load of 500 kΩ. The peak values of the output power (dashed lines) were calculated on the base of the measured values of the output voltage.

2.2.2.3 Comparison of Theory and Experiment

Earlier in the previous chapter, a mathematical model was constructed for PEGs with a height of 11 mm (as used in this section) and a solution was obtained for when the loading had an arbitrary shape. In this section, we compare the experimental data with theoretical calculations, taking as a base the initial data of the experiment, as well as the form of the loading force.

As seen in Fig. 2.18, the theoretical calculations based on the developed model demonstrate good agreement with the experimental data. The average arithmetic error does not exceed 7%.

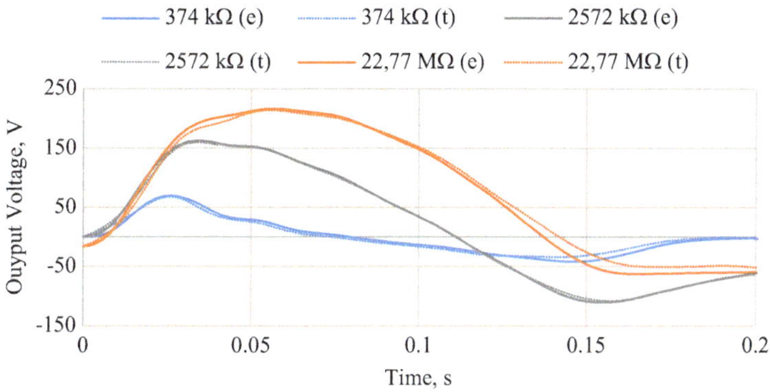

Fig. 2.18 Comparison of calculated and experimental data for PEGs of 11 mm height: t—theory (dotted line), e—experiment (solid line)

2.2.3 *Quasi-Static Loading*

Quasi-static loading is loading in which inertial phenomena are not taken into account. Time and weight can be neglected. In the field of energy harvesting, this kind of loading is interesting for constructions in which PEG is supposed to be integrated, along with dynamic loads, quasi-static loads are also present. Therefore, the evaluation of the output characteristics of PEGs operating under quasi-static loading is also of practical interest.

2.2.3.1 Description of the Test Set-up and Samples

At the first stage of the experiment, with the help of measuring instruments that were not part of the test bench, the parameters of the PEG were determined: (i) linear dimensions (using standard meters of linear quantities), (ii) electric capacitance C by using an immittance meter MNIPI E7-20, and (iii) piezomodule d_{33}, measured in quasi-static mode at a frequency of 110 Hz by using a piezoelectric d_{33} meter YE 2730A.

Further measurements of PEG characteristics in the quasi-static mode were performed on a standard MI-40KU test machine at various loading speeds from 0.9 to 4.3 kN/s. The registration and processing of the experimental data was carried out using an ADC E-14-140 M. The magnitude of the loading force acting on the PEG was recorded with a dynamometer, built into the testing machine, and the recording of the values of the electric voltage was carried out at the load resistance, connected parallel to the tested generator (see Fig. 2.19).

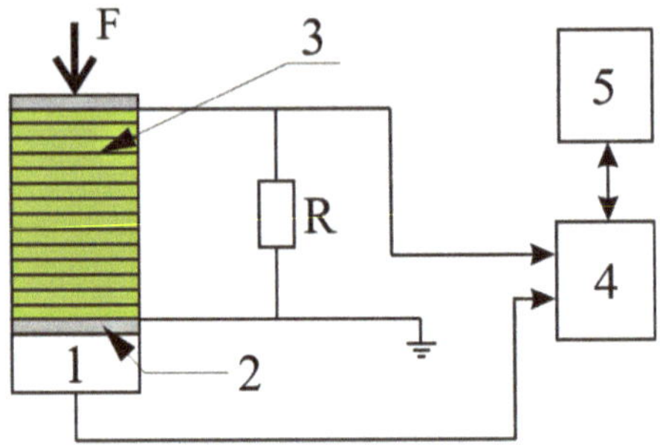

Fig. 2.19 Structural diagram of the test bench: 1—dynamometer; 2—metal overlay; 3—piezoelectric element; 4—ADC module; 5—computer; *R* is the electrical resistance

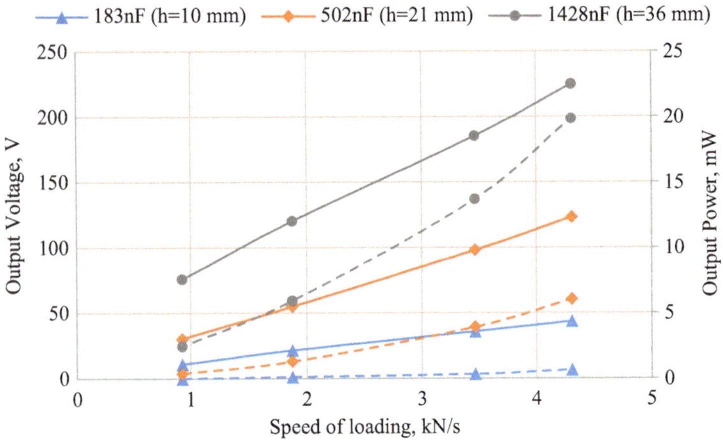

Fig. 2.20 Dependence of output voltage (solid lines) and output power (dashed lines) of generators on the speed of mechanical loading at different capacities of PEG: 26 × 16 × 36 mm³ with a capacity of 1428 nF, 24 × 16 × 21 mm³ with a capacity of 502 nF and 24 × 16 × 10 mm³ with the capacity of 183 nF

2.2.3.2 Experiment

Multilayer PEGs of prismatic geometry with a cross-sectional dimension of 24 × 16 mm² and various heights of 36, 21, and 10 mm were tested. Multilayer PEGs are made of piezoelements, based on piezoceramics PZT-19 M with a thickness of 0.5 mm, the electrodes of which are applied to the external surfaces of the PEG and connected in parallel. Sintered using traditional ceramic technology, piezoelements were polarized along height ($d_{33} = 360$ pC/N).

In accordance with the schedule of the experiment for a fixed value of electrical resistance *R* and various speeds of mechanical loading equal to 0.9; 1.9, 3.47, and 4.3 kN/s, the values of the output voltage of PEG were recorded. The constructed electric voltage diagrams are represented in Fig. 2.20 by solid lines.

As follows from Fig. 2.20, the maximum output voltage of 220 V is reached at a loading speed of 4.4 kN/s for the PEG model with a capacity of 1,428 nF. Moreover, from the plots, we can unambiguously conclude that under the quasi-static loading regime, the larger the electrical capacitance of the PEG, the greater its output voltage.

Based on the results of the measured peak values of the PEG output voltage and the corresponding values of the resistance *R*, the output power values were calculated. The constructed graphs of the dependence of the power on the rate of mechanical loading are shown in Fig. 2.20 by dashed lines.

2.3 Conclusions

In this chapter, a series of laboratory experiments was described to study the output performance of two types of PEG under various loading conditions.

For cantilever type only harmonic loading was considered. A comparison was made with numerical experiments, which showed the following. The difference between the calculated and experimental resonance frequencies did not exceed 5%. The spread between the experiment and the calculation by measuring the output voltage was 18%. This variation is due to the difference between the calculated and actual capacitance of the piezoelectric elements. Nevertheless, in the results obtained there is a qualitative agreement with the experimental data.

For the stack type, three types of loading were investigated: harmonic, pulse and quasi-static. In addition, a comparison was made with numerical experiments. So in the case of harmonic loading of one of the samples, the difference between the experimental and calculated data did not exceed 6%. While for the other the difference was 16%, which was due to the fact that a modified composition of ceramics was used for its production, for which there is not yet a complete set of material constants. Nevertheless, even with such an error, one can note the qualitative similarity of the results. For the case of impulse loading, the arithmetic mean error did not exceed 7%.

Chapter 3
Mathematical Modeling of Flexoelectric Effect

To study the flexoelectric effect, a three-point excitation model for a piezoelectric element with a proof mass at its center was selected. This type of excitation occupies an "intermediate" place between cantilever- and stack-type PEGs. The axial action of the proof mass along the center of the plate leads to an elastic deflection (see Fig. 3.1), which causes compression of the upper and tension of the lower near-electrode layers, as well as the creation (or modification of existing stationary gradients) of counter-directed gradients of deformation.

The possibility of the occurrence of polarization under the action of the strain gradient was first pointed out by Mashkevich and Tolpygo [103, 168]. They obtained a mathematical expression for the relation between the amplitudes of polarization and the gradient of deformation in an acoustic wave for structures of the diamond type. The first step in describing the phenomenon was undertaken by Kogan [84]. Some symmetry aspects of the description within the framework of the scheme, proposed in [84], were discussed in the work of Indenbom, Loginov, and Osipov [79]. In the same work, an attempt was made to microscopically describe the flexoelectric effect for the case of a static strain gradient. The effect was studied experimentally by Zheludev [185]. The experimentally-obtained value of the proportionality coefficient between the polarization and the strain gradient was four orders of magnitude larger than the rough theoretical estimate.

The purpose of this chapter is to simulate the effect of mechanical loading by using a three-point scheme of bending on the possibility of obtaining an electrical response from unpolarized ferroelectric ceramics plates.

S. N. Shevtsov et al., *Piezoelectric Actuators and Generators for Energy Harvesting*, Innovation and Discovery in Russian Science and Engineering,
https://doi.org/10.1007/978-3-319-75629-5_3

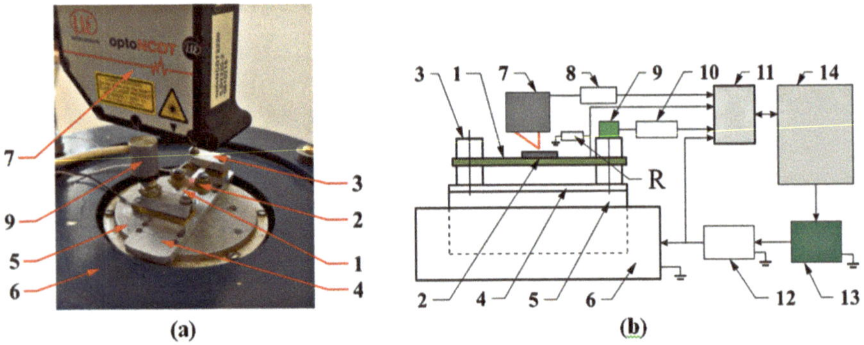

Fig. 3.1 (**a**) General view of the measuring set-up, where 1 is the piezoelectric plate, 2 is the proof mass, 3 is the first fixing point, 4 is the piezogenerator base, 5 is the vibrator working table, 6 is the electromagnetic vibrator, 7 is the optical sensor of linear displacement, and (**b**) its structural scheme, where 1 is the sample under study, 2 is the proof mass, 3 is the first fixing point, 4 is the mounting base, 5 is the shaker's moving table, 6 is an electromagnetic shaker VEBRobotron 11,077, 7 is the optical linear sensor, 8 is the optical sensor controller, 9 is the acceleration sensor ADX 103 (second fixing point), 10 is the acceleration sensor matching device, 11 is the external module of ADC/DAC E14-440D, 12 is the power amplifier, 13 is the functional signal generator, 14 is the computer, R is the load resistance

3.1 Investigation of Output Voltage in Unpolarized Ceramics

3.1.1 Samples for Study and Experimental Procedure

Four batches of plates, made from ceramics PZT-19 with the size of $50 \times 4 \times 0.7\ \text{mm}^3$, were used in this study. The electrodes from the silver were applied to the large surfaces. Before applying the electrodes, samples in batches 2, 3, and 4 were additionally damaged during processing one of the sides using grinding powders with grain sizes of 5, 10, and 20 μm, respectively. Using this method, a different depth of diffusion of Ag was obtained. After the Ag was burned into the surface, metal-ceramic layers with various thicknesses were created on opposite sides. At the boundaries of these layers in batches 2, 3 and 4, gradients of mechanical deformations, as well as tangential compressive stresses, were non-equivalent to the samples in batch 1. Compressive stresses are due to the difference in the temperature coefficients of linear expansion for Ag and ferroelectric ceramics on cooling from 730 °C after burning [158].

Figure 3.1 presents a photograph of the measuring set-up and its structural diagram

The measuring set-up used in this experiment is almost identical to the set-up described in the previous chapter, which was used to determine the output characteristics of the cantilever-type PEG. The main difference is in the device for holding the sample, which provided a cantilever clamp on both sides.

3.1.2 Results of the Experiment and Discussion

Figure 3.2 shows the dependence of the electrical response of the plates on the acceleration of the shaker in the presence of the proof mass $M = 3.1$g. The analysis of Fig. 3.2a (for plate # 3 of batch 1 – 3B1) shows that in batch 1 after Ag burning, the gradients of static deformation at the boundaries of cermet layers with a ceramic volume, as well as the tangential compressive stresses in the near-electrode layers create the electric polarization. This polarization is stable to multiple mechanical influences. This conclusion is illustrated by the curves in Fig. 3.2a, the first of which (down) was obtained before, and the second (up) after turning the plate under M on 180°. The output electrical signal from the action of M is due to the compression of the near-surface layer and the tension of the lower layer. In this batch, for each position of the plate surfaces with respect to M, the values of the output voltage were the same. This fact indicates the identity of the created polarizations for opposite faces.

Figure 3.2b demonstrates a 25% increase in the electrical response of the 6B3 sample, when the more defective face is located on the opposite side of M (down). In this case, as in batch 1, the deflection of the plate under M creates a tensile mechanical stress opposite to the existing one, which arises during the formation of the cermet layer at this surface. The difference in the magnitude of the output voltage can be provided by a higher value of polarization at this surface.

The main result of this experiment is that the described polarization originates on opposite surfaces of ceramic plates from PZT-19 when Ag is fired. The values of this polarization depend on the degree of surface roughness before the application of the electrodes. Because the effect is investigated at room temperature in the ferroelectric phase for the PZT-19, it is impossible to unambiguously interpret it as flexoelectrical

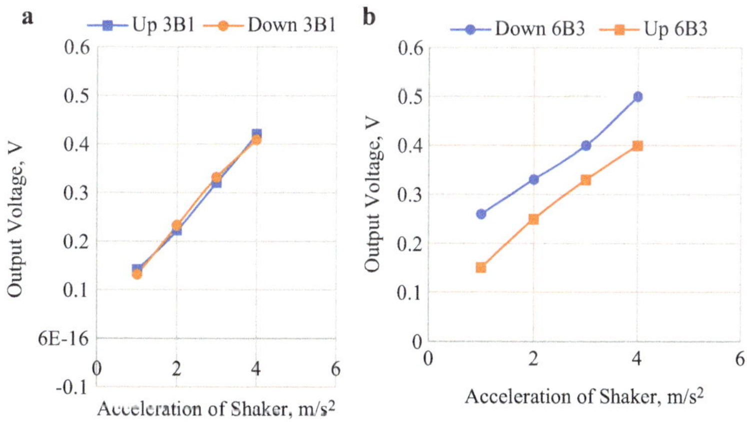

Fig. 3.2 Relationship between electrical response to mechanical impact of $M = 3.1$ g and acceleration of the worktable: before (up) and after (down) rotation of the specimens on 180° in the clamping device: beams from batches #1 (**a**) and #3 (**b**)

and “giant.” At the same time, with a high probability, it can arise in a very thin transition layer on the boundary of the cermet layer with a homogeneous array of ceramics. However, its output voltage of 0.5 V for a plate thickness of 0.7 mm without preliminary polarization in an electric field is already of practical interest for creating a multi-layer transducer with thinner plates.

3.2 Investigation of the Flexoelectric Effect in Unpolarized Ceramics

3.2.1 Formulation of the Problem for Flexoelectrical Beam

Most of the works devoted to modeling the flexoelectric effect in solids [1, 76, 100, 102] use the variational approach proposed by Mindlin [110]. We write down the variational principle for an electroelastic body as follows:

$$\begin{gathered}\delta\int_{t_1}^{t_2} dt \iiint_V \left[\frac{1}{2}\rho|\dot{\mathbf{u}}|^2 - \left(W^L - \frac{1}{2}\varepsilon_0|\nabla\varphi|^2 + \mathbf{P}\cdot\nabla\varphi\right)\right] dV + \\ \int_{t_1}^{t_2} dt \iiint_V (\mathbf{q}\cdot\delta\mathbf{u} + \mathbf{E}^0\cdot\delta\mathbf{P}) dV + \int_{t_1}^{t_2} dt \iint_S (\mathbf{t}\delta\mathbf{u}) = 0,\end{gathered} \tag{3.1}$$

where V is the volume of the body, ρ is the density, $\mathbf{u}$ is the displacement vector, W^L is the internal energy density, ε_0 is the electric constant, $\mathbf{P}$ is the polarization vector, ϕ is the electric potential, $\mathbf{q}$ is the mass forces, $\mathbf{E}^0$ is the external electric field, S is the surface in which the volume V is enclosed, $\mathbf{t}$ is the surface forces.

In order to account for the influence of the strain gradient, we will use the potential energy density in the form [148]:

$$\begin{aligned} W^L(\mathbf{P},\boldsymbol{\varepsilon},\nabla\nabla\mathbf{u}) = {} & \frac{1}{2}\mathbf{P}\cdot\mathbf{a}\cdot\mathbf{P} + \frac{1}{2}\boldsymbol{\varepsilon}:\mathbf{c}:\boldsymbol{\varepsilon} + \mathbf{P}\cdot\mathbf{d}:\mathbf{S} \\ & + \mathbf{P}\cdot\mathbf{f}:\nabla\nabla\mathbf{u} + \frac{1}{2}\nabla\nabla\mathbf{u}:\mathbf{g}:\nabla\nabla\mathbf{u}. \end{aligned} \tag{3.2}$$

where $\mathbf{a}$ is the inverse dielectric susceptibility, $\boldsymbol{\varepsilon}$ is the small deformation tensor, $\mathbf{c}$ is the tensor of elastic moduli, $\mathbf{d}$ is the tensor of piezomodules, $\mathbf{f}$ is the tensor of flexoelectric moduli, $\mathbf{g}$ is a tensor that describes pure nonlocal elastic effects, $\nabla\nabla\mathbf{u}$ is the gradient of deformation.

We substitute the expression for the potential energy Eq. (3.2) into the first term of the variational principle in Eq. (3.1):

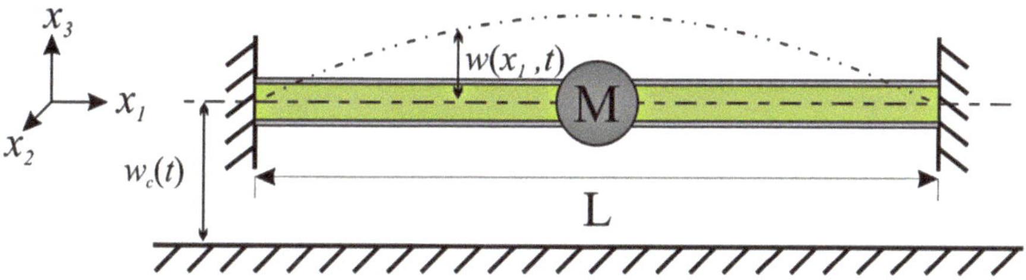

Fig. 3.3 Kinematic scheme of loading of flexoelectric beam

$$\delta\int_{t_1}^{t_2} dt \iiint_V \left[W^L - \frac{1}{2}\varepsilon_0|\nabla\varphi|^2 + \mathbf{P}\cdot\nabla\varphi\right]dV =$$
$$\int_{t_1}^{t_2} dt \iiint_V \left(\frac{\partial W^L}{\partial \mathbf{P}}\delta\mathbf{P} + \frac{\partial W^L}{\partial \boldsymbol{\varepsilon}}\delta\boldsymbol{\varepsilon} + \frac{\partial W^L}{\partial(\nabla\nabla\mathbf{u})}\delta(\nabla\nabla\mathbf{u})\right)dV \quad (3.3)$$
$$-\int_{t_1}^{t_2} dt \iiint_V (\varepsilon_0\nabla\varphi\delta(\nabla\varphi) - \mathbf{P}\delta(\nabla\varphi) - \nabla\varphi\delta\mathbf{P})dV.$$

Since the deformations in this problem are small, to simplify the solution we move from the general formulation of the problem to the one-dimensional case by adopting the Euler-Bernoulli hypotheses.

The excitation of the vibrations of the beam, presented in Fig. 3.3, occurs through the displacement of two clamps relative to a certain plane. Therefore, the absolute displacement of the beam along the coordinate x_3 will consist of the movement of $w_c(t)$ and the relative displacement of the beam $w(x_1, t)$. Taking into account the foregoing, the displacement vector $\mathbf{u}$ takes the following form:

$$\mathbf{u} = \left\{-x_3\frac{\partial w(x_1,t)}{\partial x_1}, 0, w(x_1,t) + w_c(t)\right\}^{\mathrm{T}} \quad (3.4)$$

Taking into account the introduced hypotheses, the non-zero components of the strain gradient are $\varepsilon_{11,\,1}$ and $\varepsilon_{11,\,3}$. Since the beam under consideration is sufficiently thin, we neglect $\varepsilon_{11,\,1}$. The deformation gradient $\varepsilon_{11,\,3}$ will cause the separation of centers of positive and negative charges in the unit cell of the material, thereby creating a polarization [53].

Expressions for the components of the strain tensor and the strain gradient, taking into account the introduced hypotheses, take the following form:

$$\begin{aligned}\varepsilon_{11} &= -x_3\frac{\partial^2 w}{\partial x_1^2},\\ \varepsilon_{11,3} &= -\frac{\partial^2 w}{\partial x_1^2}.\end{aligned} \tag{3.5}$$

We also take the following simplifications for the polarization vector:

$$\mathbf{P}(x_1, x_3, t) = \{0, 0, P(x_1, x_3, t)\}. \tag{3.6}$$

Taking into account the introduced simplifications and hypotheses for the convenience of writing further calculations, we take the following notation [53]:

$$a = a_{33},\ c = c_{1111},\ d = d_{311},\ f = f_{3113},\ g = g_{113113} \tag{3.7}$$

The density of potential energy takes the following form:

$$W^L = \frac{1}{2}aP^2 + \frac{1}{2}cx_3^2\left(\frac{\partial^2 w}{\partial x_1^2}\right)^2 - dx_3P\frac{\partial^2 w}{\partial x_1^2} - fP\frac{\partial^2 w}{\partial x_1^2} + \frac{1}{2}g\left(\frac{\partial^2 w}{\partial x_1^2}\right)^2. \tag{3.8}$$

Suppose that external forces and electric fields are absent. Substituting the expression for the potential energy density Eq. (3.8) into Eq. (3.3), we get:

$$\begin{aligned}&\int_{t_1}^{t_2} dt \iiint_V \rho(\ddot{w} - \ddot{w}_c)\delta w dV + \int_{t_1}^{t_2} dt \iiint_V \left[\left(aP - dx_3\frac{\partial^2 w}{\partial x_1^2} - f\frac{\partial^2 w}{\partial x_1^2} + \frac{\partial \varphi}{\partial x_3}\right)\delta P + \right.\\ &\quad + \left(cx_3^2\frac{\partial^2 w}{\partial x_1^2} - dx_3P - fP + g\frac{\partial^2 w}{\partial x_1^2}\right)\delta\left(\frac{\partial^2 w}{\partial x_1^2}\right) + \\ &\quad\quad \left. + \left(P - \varepsilon_0\frac{\partial \varphi}{\partial x_3}\right)\delta\left(\frac{\partial \varphi}{\partial x_3}\right) + \left(-\varepsilon_0\frac{\partial \varphi}{\partial x_1}\right)\delta\left(\frac{\partial \varphi}{\partial x_1}\right)\right] dV = 0.\end{aligned} \tag{3.9}$$

Since the variation of the polarization δP is arbitrary, then:

$$P = \frac{1}{a}\left(-dx_3\frac{\partial^2 w}{\partial x_1^2} - f\frac{\partial^2 w}{\partial x_1^2} + \frac{\partial \varphi}{\partial x_3}\right) \tag{3.10}$$

After substitution of Eq. (3.10) into Eq. (3.9) and integrating over the cross-sectional area S, we get:

$$\int_{t_1}^{t_2} dt \int_0^L \rho \mathbf{I}_1 (\ddot{w} - \ddot{w}_c) \delta w dx_1 + \int_{t_1}^{t_2} dt \int_0^L \{ \left[\left(c - \frac{d^2}{a} \right) \mathbf{I}_3 - \frac{2df}{a} \mathbf{I}_2 - \left(\frac{f^2}{a} - g \right) \mathbf{I}_1 \right] \frac{\partial^2 w}{\partial x_1^2}$$
$$+ \iint_S \left(\frac{d}{a} x_3 + \frac{f}{a} \right) \frac{\partial \varphi}{\partial x_3} ds \} \delta \left(\frac{\partial^2 w}{\partial x_1^2} \right) dx_1 = 0, \tag{3.11}$$

where:

$$\mathbf{I}_1 = \iint_S dS, \ \mathbf{I}_2 = \iint_S x_3 dS, \ \mathbf{I}_3 = \iint_S x_3^2 dS.$$

We introduce the notation for the right factor in the first term of the integrand of the second integral in Eq. (3.11):

$$\mathrm{EI}^* = \left(c - \frac{d^2}{a} \right) \mathbf{I}_3 - \frac{2df}{a} \mathbf{I}_2 - \left(\frac{f^2}{a} - g \right) \mathbf{I}_1. \tag{3.12}$$

The coefficient EI^* can be interpreted as the effective flexural rigidity of a flexoelectric beam.

We will assume that the electric field is linear in the thickness of the beam, then:

$$E_3 = -\frac{\partial \varphi}{\partial x_3} = \mathrm{const} = -\frac{v(t)}{h}, \tag{3.13}$$

where $v(t)$ is the electric potential between two electrodes on large surfaces of a beam, h is the thickness of a beam.

Since the measurement of the electrical voltage occurs at an electrical resistance R, the current flowing through the resistor will be equal to the time derivative of the averaged electric displacement D_3:

$$i(t) = \frac{v(t)}{R} = \frac{d}{dt} \left(\frac{1}{h} \iiint_V D_3 dV \right), \tag{3.14}$$

where $D_3 = -\varepsilon_0 \nabla \phi + P$. Taking into account Eq. (3.10), we obtain the equation of an electrical circuit with a flexoelectric coupling:

$$\frac{v(t)}{R} = -\frac{BL}{h} \left(\varepsilon_0 + \frac{1}{a} \right) \dot{v}(t) + \frac{1}{h} \int_0^L \left(\frac{d}{a} \mathbf{I}_2 + \frac{f}{a} \mathbf{I}_1 \right) \frac{\partial^2 \dot{w}}{\partial x_1^2} dx_1. \tag{3.15}$$

To solve the problem of forced vibrations of a flexoelectric beam, we will use the Kantorovich method [81]. We represent the relative displacements of a beam as a series expansion:

$$w(x_1, t) = \sum_{i=1}^{N} \eta_i(t)\phi_i(x_1), \tag{3.16}$$

where N is the number of vibration modes under consideration, $\eta_i(t)$ are unknown generalized coordinates, $\varphi_i(x_1)$ are known trial functions that satisfy the boundary conditions.

Substitute Eq. (3.13) and Eq. (3.16) in Eq. (3.11) and Eq. (3.15). In Eq. (3.11) we equate to zero the coefficients before the independent variation $\delta\eta$. We have obtained a system of two differential equations describing the forced oscillations of a flexoelectric beam, connected to a resistor:

$$\begin{aligned} &\mathbf{M}\ddot{\boldsymbol{\eta}}(t) + \mathbf{D}\dot{\boldsymbol{\eta}}\,(t) + \mathbf{K}\boldsymbol{\eta}(t) - \boldsymbol{\Theta}v(t) = \mathbf{p} \\ &C_f\dot{v}\,(t) + \boldsymbol{\Theta}^{\mathrm{T}}\dot{\boldsymbol{\eta}}\,(t) + \frac{v(t)}{R} = 0 \end{aligned} \tag{3.17}$$

here:

$$\begin{aligned} C_f &= \frac{bL}{h}\left(\varepsilon_0 + \frac{1}{a}\right), \\ M_{ij} &= \int_0^L \rho\,\mathbf{I}_1\phi_i(x_1)\phi_j(x_1)dx_1, \\ K_{ij} &= \int_0^L \mathrm{EI}^*\phi_i''(x_1)\phi_j''(x_1)dx_1, \\ p_i &= \int_0^L \ddot{w}_c(t)\rho\,\mathbf{I}_1\phi_i\,dx_1, \\ \Theta_i &= \int_0^L \frac{\phi_i''\,dx_1(x_1)}{h}\left(\frac{d}{a}\mathbf{I}_2 + \frac{f}{a}\mathbf{I}_1\right)dx_1, \end{aligned} \tag{3.18}$$

where C_f is the effective capacity, b, h, and L are the width, height, and length of the piezoelectric element.

Now it remains to solve the problem of finding trial functions to satisfy the boundary conditions.

3.2.2 Boundary Conditions

To find $\varphi_i(x_1)$, we solve the problem of free vibrations for a beam clamped at both ends and having a proof mass at the center (see Fig. 3.1).

To take into account the influence of the proof mass M, located in the middle of the beam, we represent the function $\varphi_i(x_1)$ in the following form:

$$\phi_i(x_1) = \begin{cases} \phi_i^{(1)}(x_1), x_1 \le \frac{L}{2} \\ \phi_i^{(2)}(x_1), x_1 > \frac{L}{2} \end{cases} \tag{3.19}$$

where $\phi_i^{(1)}(x_1)$ corresponds to the shape of the left half of the beam, and $\phi_i^{(2)}(x_1)$ corresponds to the right half.

In addition, it is necessary to take into account the influence of Min the system in Eq. (3.17). To do this, we add to Eq. (3.18) two components responsible for the influence of the mass M:

$$\begin{aligned} M_{ij} &= \int_0^L \rho \mathbf{I}_1 \phi_i(x_1)\phi_j(x_1)dx_1 + M\phi_i\left(\frac{L}{2}\right)\phi_j\left(\frac{L}{2}\right), \\ p_i &= \int_0^L \ddot{w}_c(t)\rho \mathbf{I}_1 \phi_i dx_1 + M\phi_i\left(\frac{L}{2}\right). \end{aligned} \tag{3.20}$$

We write the solution in common form for each part of the beam:

$$\begin{aligned} \phi_i^{(1)}(x_1) &= a_{1,i}\sin(\beta_i x_1) + a_{2,i}\cos(\beta_i x_1) + a_{3,i}\mathrm{sinh}(\beta_i x_1) + a_{4,i}\mathrm{cosh}(\beta_i x_1), \\ \phi_i^{(2)}(x_1) &= a_{5,i}\sin(\beta_i x_1) + a_{6,i}\cos(\beta_i x_1) + a_{7,i}\mathrm{sinh}(\beta_i x_1) + a_{8,i}\mathrm{cosh}(\beta_i x_1). \end{aligned} \tag{3.20}$$

We write out the boundary conditions for the ends of the beam and the coupling condition at the center of the beam:

$$\begin{matrix} & \phi_i^{(1)}\left(\frac{L}{2}\right) = \phi_i^{(2)}\left(\frac{L}{2}\right), & \\ \phi_i^{(1)}(0) = 0, & \phi_i^{(1)}\left(\frac{L}{2}\right) = \phi_i^{(2)\prime}\left(\frac{L}{2}\right), & \phi_i^{(2)}(L) = 0, \\ \phi_i^{(1)\prime}(0) = 0, & \phi_i^{(1)\prime\prime}\left(\frac{L}{2}\right) = \phi_i^{(2)\prime\prime}\left(\frac{L}{2}\right), & \phi_i^{(2)\prime}(L) = 0, \\ & \phi_i^{(1)\prime\prime\prime}\left(\frac{L}{2}\right) = \phi_i^{(2)\prime\prime\prime}\left(\frac{L}{2}\right) - \alpha\beta^4\phi_i^{(1)}\left(\frac{L}{2}\right), & \end{matrix} \tag{3.21}$$

where $\alpha = M/\rho \mathbf{I}_1 L$

Satisfying the boundary conditions, we obtain a homogeneous system of eight equations with eight unknowns:

$$\Lambda = \begin{pmatrix} a_{1,1} & \cdots & a_{1,8} \\ \vdots & \ddots & \vdots \\ a_{8,1} & \cdots & a_{8,8} \end{pmatrix} \tag{3.22}$$

We need to find the determinant of this system in order to find the eigenvaluesβ_i. Since $\det(\Lambda) = 0$ is a transcendental equation, we will search for its solution using numerical methods. Having obtained the set of β_i, we calculate the coefficients a_i for the required number of modes of oscillations N.

3.2.3 Solution

Since we will consider the case of harmonic excitation of the base, the stages of finding the solution will be similar to those given in Chap. 1 for cantilever-type PEGs. The solution for the system of equations (Eq. 3.17) has the form:

$$\begin{aligned} \tilde{\boldsymbol{\eta}} &= \left[-\omega^2\mathbf{M} + i\omega(\mu\mathbf{M} + \gamma\mathbf{K}) + \mathbf{K} + \frac{i\omega\boldsymbol{\Theta}\boldsymbol{\Theta}^T}{i\omega C_f + \frac{1}{R}}\right]^{-1}\tilde{\mathbf{p}} \\ \tilde{v} &= -\frac{i\omega\boldsymbol{\Theta}^T}{i\omega C_f + \frac{1}{R}}\left[-\omega^2\mathbf{M} + i\omega(\mu\mathbf{M} + \gamma\mathbf{K}) + \mathbf{K} + \frac{i\omega\boldsymbol{\Theta}\boldsymbol{\Theta}^T}{i\omega C_f + \frac{1}{R}}\right]^{-1}\tilde{\mathbf{p}} \end{aligned} \tag{3.23}$$

3.2.4 Numerical Experiment

As input parameters of the model, we use the initial data from the experiment. We will consider a piezoceramic beam made of unpolarized ceramic PZT-19, which has the geometric and physical properties given in Table 3.1 [24, 94, 98].

Since unpolarized ceramics is a material with central symmetry, the piezoelectric modulus d of such a material will be equal to zero. The inverse dielectric susceptibility was calculated using the formula $a = (\varepsilon\varepsilon_0 - \varepsilon_0)^{-1}$, and the flexoelectric module is $f = -a\mu_{12}$. The amplitude of the displacement of the base is $\tilde{w}_c =$ 0.03 mm. The coefficients of the modal damping are $\xi_1 = \xi_2 = 0.02$.

Let us construct the amplitude-frequency characteristics of the motion of the middle of the beam and the electric voltage for various load resistances. The calculated resonance frequency is 504 Hz.

Figure 3.4 shows that the maximum displacement at resonance is 1 mm.

The values of the electrical voltage, shown in Fig. 3.5, differ from those obtained in the experiment described in the beginning of this chapter.

Since the definition of material constants affecting higher-order effects (such as the flexoelectric effect) is a complex and not completely solved research problem, in our case we can try to vary the coefficient μ_{12}.

During the variation of the coefficient, it was found that when a certain value of the coefficient is reached, the effective flexural rigidity EI^* becomes negative.

Table 3.1 Parameters of the ferroelectric ceramic beam

	Piezoelement
Geometrical dimensions ($L_0 \times b \times h$)	$50 \times 4 \times 0.7$ mm^3
Length of working part of the specimen (L)	35 mm
Density (ρ)	7,280 kg/m^3
Elastic modulus (c)	114.8 GPa
Elastic compliance $\left(s_{11}^E\right)$	17.5×10^{-12} Pa
Relative permittivity $\left(\varepsilon_{33}^S/\varepsilon_0\right)$	682.6
Inverse dielectric susceptibility (a)	0.166 GNm^2/C^2
Flexoelectrical factor (μ_{12})	2 μC/m
Flexoelectrical modulus (f)	−331 Nm/C
Higher order elastic modulus (g)	1.75 μN

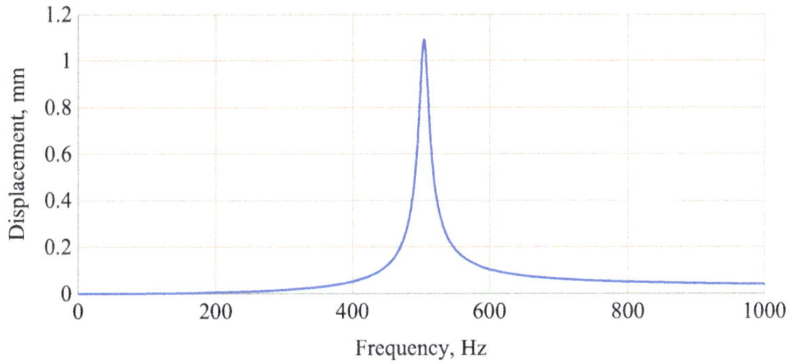

Fig. 3.4 Frequency response of the beam's center displacement, obtained from the numerical experiment

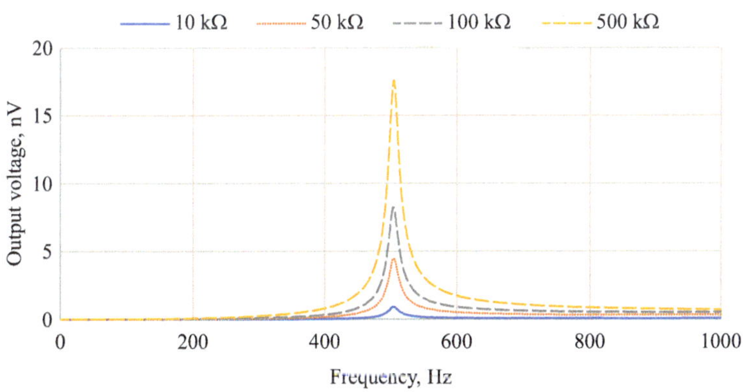

Fig. 3.5 Frequency response of output voltage across the resistor with different values of the load resistance, obtained from the numerical experiment

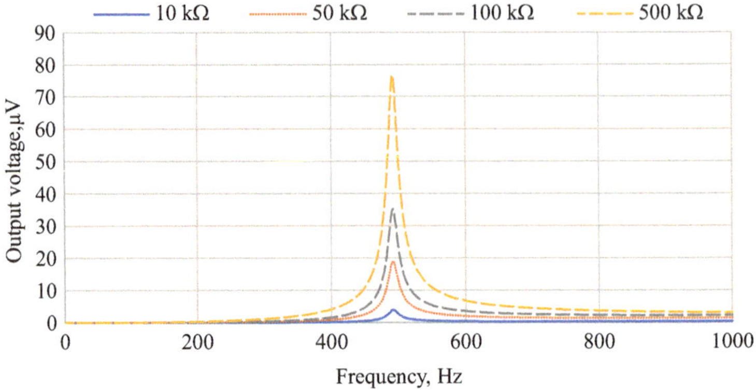

Fig. 3.6 Frequency response of the voltage across the resistor with different values of the load resistance, obtained from the numerical experiment

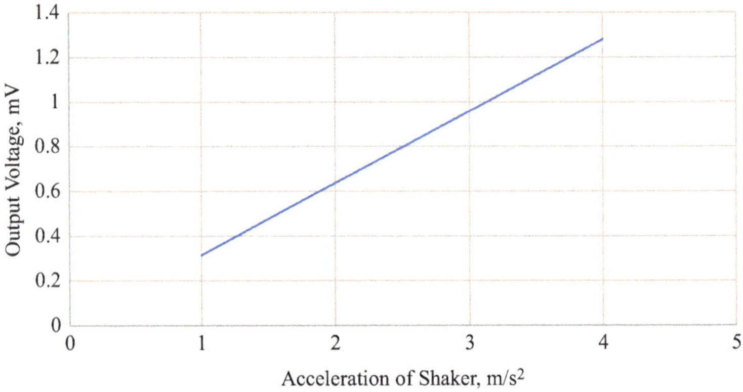

Fig. 3.7 Dependence of the electrical response on the acceleration of the shaker, measured at the resonance frequency at $R = 360$ kΩ and $M = 3.1$ g

Therefore, we chose the nearest value in the vicinity of the transition point, equal to 10^{-3}, and built the amplitude-frequency characteristic of the electric voltage.

Figure 3.6 shows that the values of the electrical voltage increased by three orders of magnitude. Nevertheless, these values are sufficiently small in comparison with the experimental values. In addition, the resonance frequency slightly decreased to 491 Hz. The amplitude of the displacement of the middle of the beam remained unchanged. Such a difference may indicate inaccurate input data, as well as the nonlinearity of the phenomenon under study.

At the beginning of the chapter in Fig. 3.2, the dependencies of the output electric voltage on the table acceleration are presented. Let us construct an analogous dependence.

The values of the output voltage presented in Fig. 3.7 are quantitatively different from the experimental values, nevertheless, they qualitatively reflect the dependence of the electric voltage on acceleration. Such a difference may indicate inaccurate input data, as well as the nonlinearity of the phenomenon under study.

The results obtained in the course of the numerical experiment showed the possibility of the appearance of an electric potential in an unpolarized piezoceramic beam and its qualitative characteristics.

3.3 Conclusion

The main result of this chapter is the formulation of the problem of forced oscillations of an unpolarized ferroelectric beam in the presence of proof mass, taking into account the flexoelectric effect.

It is shown that the output electric potential can arise in unpolarized samples, and its values can serve to determine the flexoelectric constants.

It is found that the variation of such a constant is possible only up to a certain point, when the effective stiffness EI becomes negative, and the results lose physical meaning.

Chapter 4
Amplified High-Stroke Flextensional PZT Actuator for Rotorcraft Application

4.1 Introduction

The vibrations and noise experienced in a helicopter are mainly caused by periodic forces generated in forward flight by the main rotor due to blade vortex interaction (BVI) and high Mach numbers at the advancing blade. These vibrations are transferred through the rotor hub and the gearbox into the fuselage and this limits the operator's capability, performance, reliability, handling qualities, and the helicopter's efficiency. Due to these negative outcomes, the active control of helicopter rotor blades has raised significant interest in the last 30 years. Many attempts to solve the problem of noise and vibrations associated with active control of the rotor blades have been made.

The first theoretical studies for rotor vibration reduction involved higher harmonic control (HHC), which is based on actuators located below the swashplate, enforcing fixed-frame oscillations at frequencies $k\Omega$, $k = 1\,..\,n_b$, where Ω is the rotor angular frequency rotor angular frequency, and n_b is the total number of blades [82, 96]. A different solution to HHC is Individual Blade Control (IBC), which is based on actuators in the rotating frame. These actuators independently change the aerodynamic properties of each blade in the real time [82, 89, 96, 108]. Both the HHC and IBC concepts allow us to reduce vibration and blade-vortex interaction (BVI)-induced noise. IBC systems are more suitable for simultaneous vibration and noise reduction, shaft power reduction and a flight envelope extension. These technologies work with a higher harmonic excitation of the blade pitch at the blade root, which has a very energy consumption. A higher harmonic excitation can also lead to blade excitation on a first twisting vibration mode [75]. With an active trailing edge (ATE) flap, the excitation is localized at a distance of 75–90% of the blade span from the rotation axis [90, 149, 150, 179]. The ATE concept can be implemented in the form of a turned discrete trailing edge flap (see Fig. 4.1) [90, 129, 141, 149, 150, 179] or as flexible locally morphing airfoil [32, 48, 108]. The actively

S. N. Shevtsov et al., *Piezoelectric Actuators and Generators for Energy Harvesting*, Innovation and Discovery in Russian Science and Engineering,
https://doi.org/10.1007/978-3-319-75629-5_4

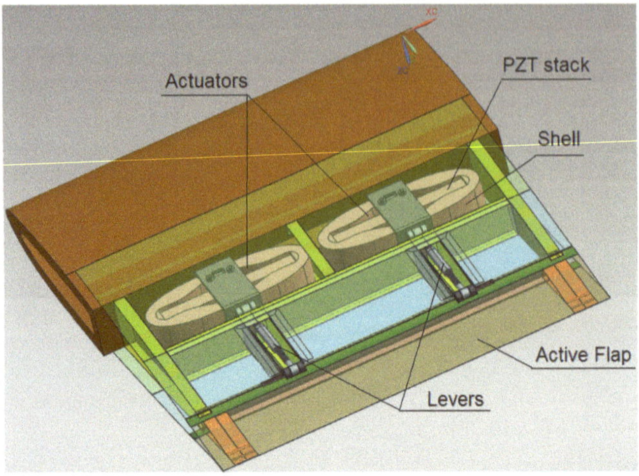

Fig. 4.1 Schematic view of part of an ATE rotor blade with two amplified flextensional actuators [156]

controlled flaps are generally 15% of a chord length and most structures are driven by using the power piezoelectric transducers. The power required by the trailing-edge flap is an important parameter that is necessary to consider for practical implementation [83, 99, 165, 179]. Generally, the most important requirements for the piezoelectric actuator for the active flap design are formulated in [108, 141, 179] as follows:

(i) high force and large displacement of actuators should be provided in compact sizes, force actuation must be able to react operational hinge moments, and stroke actuation must be capable of $\pm 5^{\circ}$ of flap motion;
(ii) high resolution within the micrometer range and very short response time below 1 ms are necessary to effectively operate under control of adaptive open-loop and closed-loop regulator at the higher harmonics (> 5/rev) [112];
(iii) the actuation mechanism of an active blade must either be protected against or withstand these forces and the large strains of the blade structure, and lifetime must exceed 10^{10} cycles;
(iv) low voltage supply below 250 V DC is preferable; low power consumption when static;
(v) actuator must be able to operate at severe flight and environment conditions (broad temperature and moisture ranges).

Due to the high operating forces and frequencies, the piezoelectric actuators are well-adapted to drive the trailing edge flaps, but relatively large displacements require some sort of mechanical amplification of the movement from these devices. Because of the very small displacements created by the piezoelectric devices of some different designs, we propose amplifying the stroke of PZT actuators. Among these designs are the following well-known actuators: X-frame [90, 141], "Diamond"

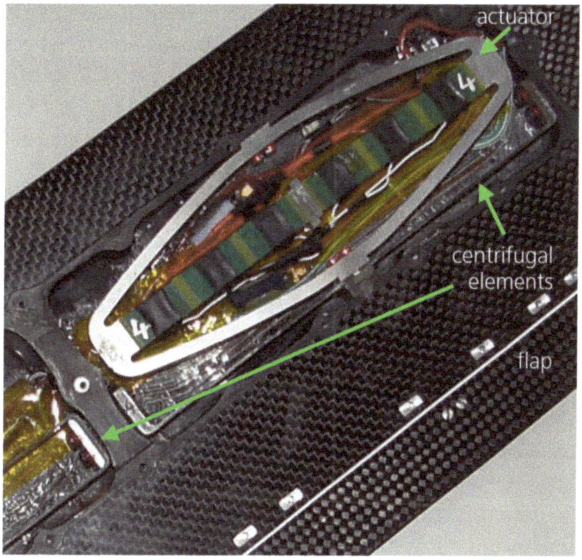

Fig. 4.2 Blade with flap-driving system [99]

[141], with lever-based amplification [129, 141], and actuators with a pre-stressed piezoelectric stack, which is located along the major axis of elliptic shell made from metal or composite materials (see Fig. 4.2) [75, 80, 83, 99, 111, 112, 165]. In this last example (sometimes called the amplified flextensional actuator) the elliptic shell supplies the stroke amplification. These actuators provide a relatively large displacement. In order to eliminate an impairment of blade balancing by actuator mass and its drive unit, carbon or glass fiber composites are used instead of metal for the shell frame of these flextensional actuators [80, 108, 111, 155, 156]. These actuators are developed and manufactured by the French companies Noliac and Cedrat Technologies (see Fig. 4.2). During operation, the aerodynamic forces are transmitted to the actuator through the levers. These forces deflect the flap in opposite direction to its active deflection. Because of acting aerodynamic forces, the higher flap deflections require larger and therefore heavier actuators, but additional mass is necessary to equilibrate the center of mass of the blade close to active flap mounting.

In our previous works [155, 156], we reported the technique for the performance optimization of the considered flextensional actuator. This technique assumed the maximum stroke at the given external force as the objective function. The dimensional weight limits, permissible operating voltage, all parameters of the given piezoelectric material, and curvature of the outer generatrix were considered as the constraints. The shape of the shell has been described by the rational Bezier curves that are defined by the coordinates of the control points with the corresponding weights. These coordinates and weights are considered as the design variables. Because the total number of degrees of freedom for this problem is very big, a genetic algorithm implemented in MATLAB© (Genetic Algorithm Toolbox) was used to change these design variables and one-quarter of actuator's 2D FEM model on each iteration step was considered. After a short explanation of the optimization approach used, we present the optimized actuator's design with some simulation results obtained in the dynamic analysis of the optimized actuator, which are compared with the experimental data. Degradation of the actuator's performance, which was observed at high frequencies and loading forces, is discussed in connection with deficient stiffness of piezoelectric ceramic, adhesive interlayers, and low output current limit of high-voltage amplifiers.

4.2 Modeling and Numerical Optimization of the Actuator Shell

The designed actuator should be used with the blades of the middle weight helicopter. Its dimensions, weight, and force parameters will be sufficiently different from those for the light helicopters studied in [75, 96, 99, 108]. Due to the structural and weight constraints of the blade structure, only the actuators that have the overall dimensions (172 $\times$ 64 $\times$ 24) mm with the elliptic shell, made from polymeric composite material, have been considered. We adopted the glass fiber epoxy polymeric composite with the longitudinal Young modulus 3×10^{10} Pa and density 1,850 kg/m^3. The size of the finished shell should provide a pre-stress for the assembled piezoelectric stack for a double-side acting actuator. The stack was constructed of a multilayer piezoelectric ceramic PZT-5H, and the thickness of each layer, polarized along the thickness, was 0.5 mm. Driving electric potential for each PZT layer electrically connected in parallel was taken up to 500 V. Geometry of a preliminary studied FEM model is presented in Fig. 4.3. At the fixed dimensional, mechanical, and electric properties of PZT stack, the

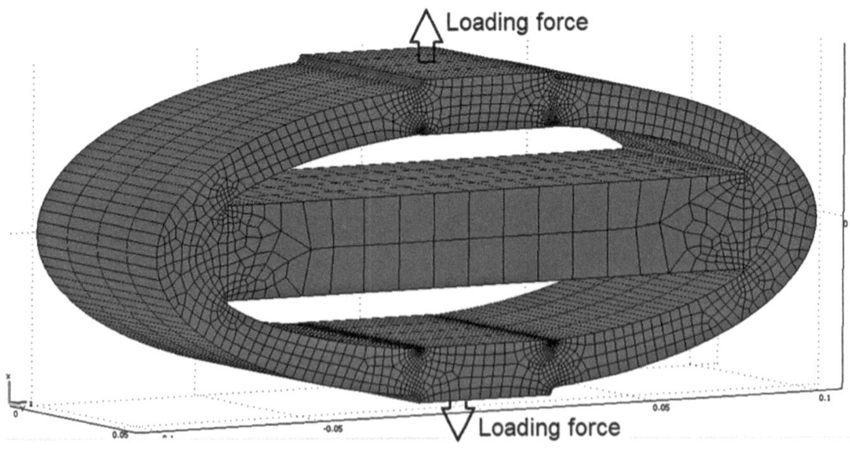

Fig. 4.3 Finite element model for preliminary actuator analysis

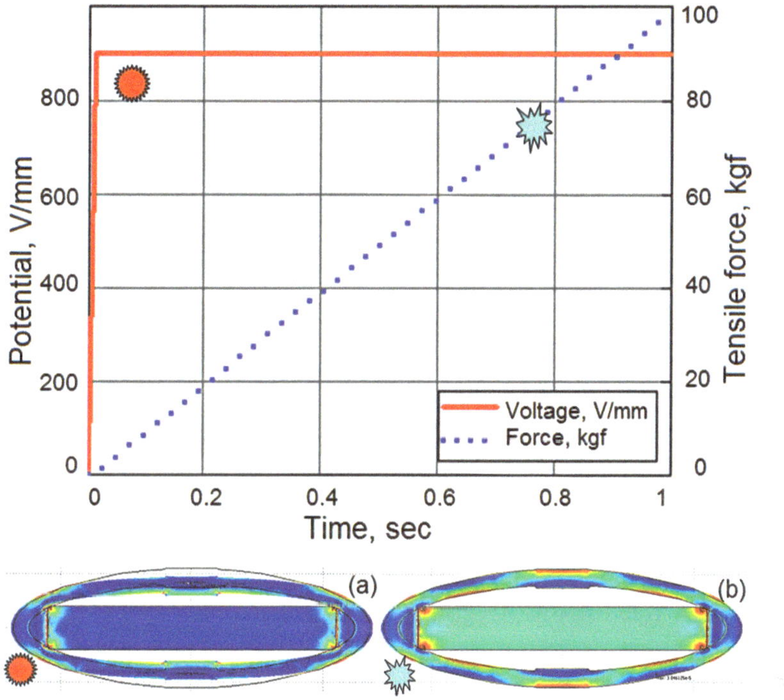

Fig. 4.4 Plot of numerical test scenario and shapes of the shell before (**a**) and after applying the external tensile force (**b**)

dependencies of the actuator's operational parameters on the thickness of the elliptic shell and its axes ratio have been studied.

The preliminary finite element (FE) analysis was performed in a static mode as follows: After application of the driving potential, the stack expands, deforming the shell and causing it to contract in the vertical direction (see Fig. 4.4). Then gradually increasing tensile stress was applied to the executive surfaces of the shell (small quadrilateral planes). When the structure was under loading, the values of operating stroke, applied reaction force, and deformation of piezoelectric stack were monitored. As soon as the displacement of the executive surfaces returned to zero, the blocking forces were recorded.

As might be expected, a thicker shell resulted in more stiffness, but less free stroke. It has also been established that shallower shells provided greater stroke amplification, but a lowered ability to counteract external loads. The total compliance of actuator was determined by the shell and not the PZT stack, which had approximately two times greater stiffness. Hence, structural optimization of the shell is necessary to enhance its stiffness without significant loss of stroke.

The basic element of the optimization process is to choose the design parameterization [25, 77]. The optimization of the shell has been fulfilled by varying the shape of the generatrix and thickness distribution of the shell. As a performance criterion

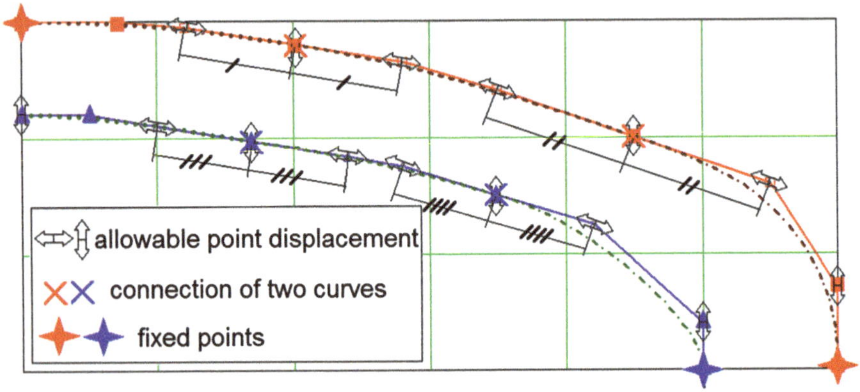

Fig. 4.5 Representation of the shell's profile by the third-order rational Bezier curves; all positions of control points are constrained by the system of inequalities

(the objective function), the operating stroke h_{op} has been chosen at the given external load F_{act}.

To describe the shape of the generatrix with the necessary flexibility, rational Bezier curves have been used. Due to symmetry of the shell, only one-quarter has been modeled and optimized. The symmetrical shell geometry is present as two branches, each of which consists of three third-order Bezier curves (see Fig. 4.5). The rational Bezier curve of nth order defined by the $(n + 1)$ control points P^i is described by the weighted sum:

$$R(u) = \sum_{i=0}^{n} w_i B_n^i(u) P^i / \sum_{i=0}^{n} w_i B_n^i(u), \tag{4.1}$$

where $B_n^i(u)$ are the Bernstein polynomials, defined as:

$$B_n^i(u) = \frac{n!}{i!(n-i)!} u^i (1-u)^{n-i}, \tag{4.2}$$

$w_i, i = 0, 1 \ldots n$ are the weights of the control points, and u is the parameter, which runs through the values from 0 to 1.

In the optimization process, only three points of the shell (marked with asterisks in Fig. 4.6) remain unchanged, so only the shape of the generatrices is optimized. In order to supply C^1 continuity at the connection of the points (denoted by $\times$), the additional restricting conditions on the location of the control points, which are adjacent to the connection point of two curves, were imposed. If P^c are the coordinates of the connection point, P^{lc} is the vector coordinates of the point adjacent to the connection with left hand, then coordinates of the point adjacent to the connection with right hand are:

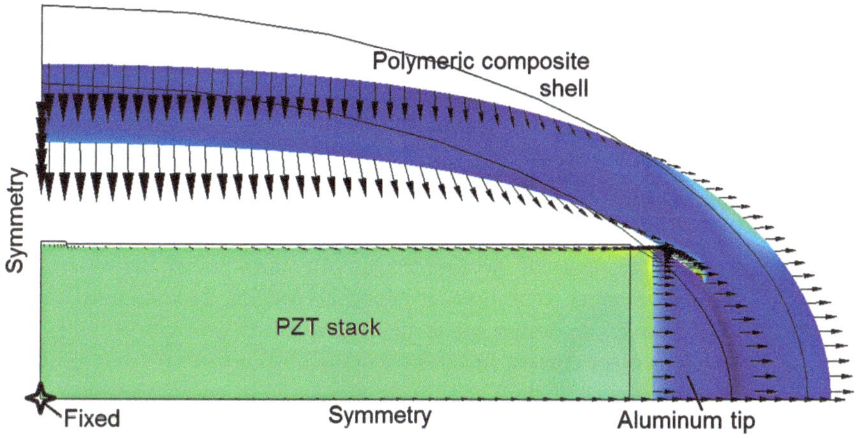

Fig. 4.6 FE model of one-quarter of actuator

$$P^{cr} = 2P^{c} - P^{lc}. \tag{4.3}$$

Figure 4.5 shows the equality of the adjacent edges of the control polygon. The disposition of points on the vertical (horizontal) lines nearby the terminal points (denoted as $\times$) of the generatrices allows one to automatically satisfy the condition of C^1 continuity due to symmetry of the shell (see Fig. 4.6). So, we have 21 degrees of freedom (DoFs) for the coordinates of the points, and 18 DoFs for the weights, with the total number of DoFs equal to 39:

$$\begin{cases} X_i > X_{(i-1)}; & i \in [1;(n-1)] \\ Y_i < Y_{(i-1)}; & i \in [2;n] \end{cases} \tag{4.4}$$

and for connection points, we have the equations:

$$\begin{cases} X_0 = 0; & Y_0 = b^{\text{in,out}} \\ Y_1 = b^{\text{in,out}} & \\ X_n = a^{\text{in,out}}; & Y_0 = 0 \\ X_{n-1} = b^{\text{in,out}} & \end{cases} \tag{4.5}$$

where the dimensions of the external generatrix a^{out}, b^{out} are fixed, but the dimensions of the internal generatrix are given by the inequalities:

$$\begin{cases} a^{\text{out}} > a^{\text{in}} = fixed \\ b^{\text{out}} > b^{\text{in}} = \text{vari}ed \end{cases} \tag{4.6}$$

All weights in Eq. (4.1) were constrained by the system of inequalities. Each of the three Bezier curves that form the generatrices is described by its own equation (Eq. (4.1)). However, to ensure the smoothness of generatrix, the end point of one curve and the initial point of the connecting curve have the same weight. So, we have the following system of constraints for the weights:

$$\begin{cases} w_0 = 1 \\ 0 < w_i < 5; \quad i \in [1; n] \end{cases} \tag{4.7}$$

To narrow down the range of the coordinates of control points and reduce the search area, some additional restrictions are introduced [155]. In particular, due to constraints by the whole flap design and technological difficulties with winding and molding the shell, its outer contour has been restricted by curves with only positive curvature. Contact surfaces of shell and D-like aluminum inserts, which should be installed between the PZT stack and the shell, are strongly assigned as cylindrical with $R = 10.5$ mm (see Fig. 4.6). Such a restriction is very important for a problem with a large number of DoFs. To solve this optimization problem, the Genetic Algorithm Toolbox MATLAB© (GA Toolbox) was used as it has the advanced optimization means and direct access to the FE computation.

All material properties and displacements were considered as linear. Quadrilateral FEM mesh consisted of about 500 elements. The FEM model operated in the Structural Mechanics—Piezoelectric mode. At each iteration step, the GA Toolbox rebuilt the shell geometry, re-meshed it, and performed the static analysis. The calculated value of the actuator stroke was transferred to the GA Toolbox, which changed the value of the design variables according to the constraints Eqs. (4.4)–(4.7). All computation flowcharts were controlled by the GA Toolbox, which in turn referred to the developed program modules, which performed the finite element analysis. These modules are standard MATLAB's m-files. The main GA Toolbox settings were: population size—20, elite count—4, crossover—scattered, mutation—adaptive feasible, hybrid function—"fminsearch".

The interesting result of this optimization is that the most efficient design is very similar to the four-bar mechanism, whose four stiff bars are connected by the revolute joints in the form of parallelogram (see Fig. 4.7, a). These rods are thicker than the shell's wall with high bending and tensile stiffness. This analogy allows us to present a simple analysis of actuator's amplification factor dependence on the ratio between big (a) to small (b) semi-axes of the pseudo-elliptical shell. By assuming the lengths of all bars are identical and constant, the dependence between changes of the small and big semi-axes lengths can be expressed in the form of the derivative:

$$db/da(a) = -1/\sqrt{c^2/a^2 - 1}, \tag{4.8}$$

which implies that amplification factor increases rapidly at increases in the ratio a/b.

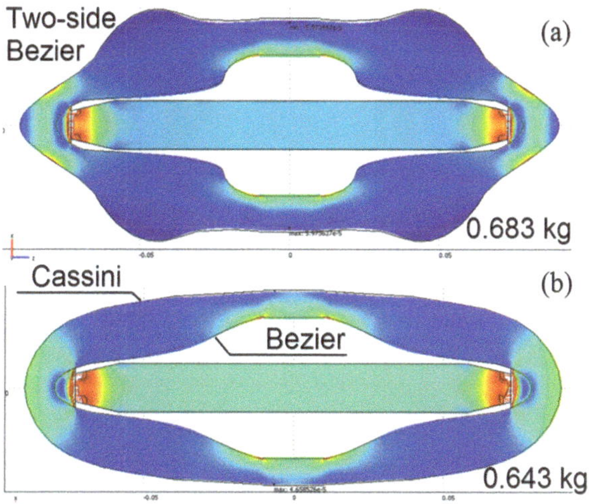

Fig. 4.7 Geometry of two FE shell models: (**a**) with the best combination "stiffness—stroke amplification", but difficult to manufacture; (**b**) with accepted "stiffness—stroke amplification" and convenient technology

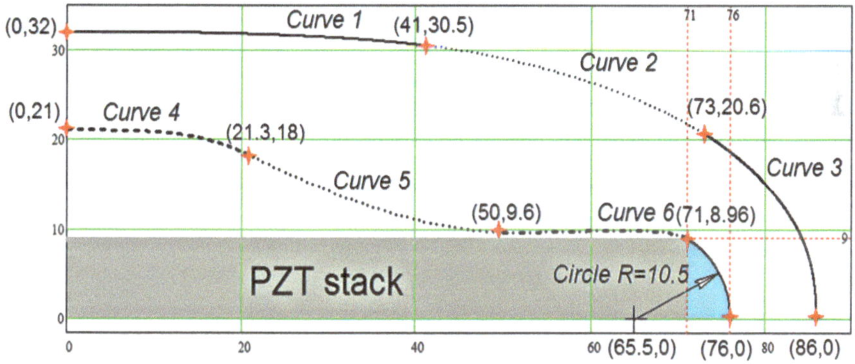

Fig. 4.8 Representation of the shell's profiles by third-order rational Bezier curves with optimized parameters (inner profile) and by Cassini oval (outer surface)

4.3 Actuator Design and Manufacture

For the convenience of manufacturing using closed-mold technology, the shell geometry was compiled from the Cassini oval (outer surface) and a combination of Bezier curves (inner surface) (see Figs. 4.7a and 4.8). Since the external profile of the shell was fully determined, the number of DoFs, which determine the inner profile (curves 4, 5, and 6) was noticeably reduced. At these conditions, the inner profile of the shell has been re-optimized.

This composite shell has been manufactured by winding of the unidirectional high-strength glass-fiber tape onto mandrel with geometry determined after

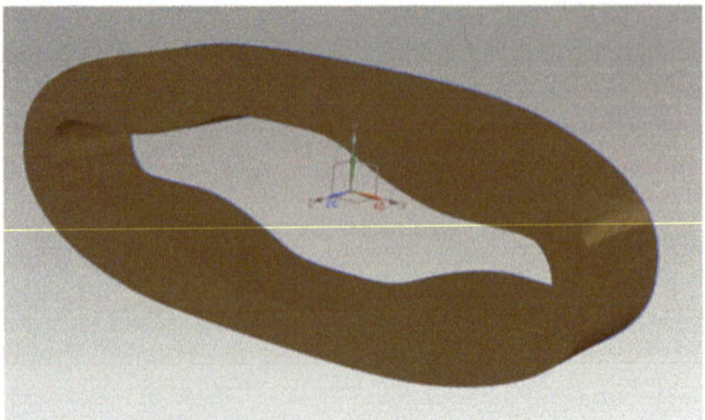

Fig. 4.9 CAD model of the actuator's composite shell

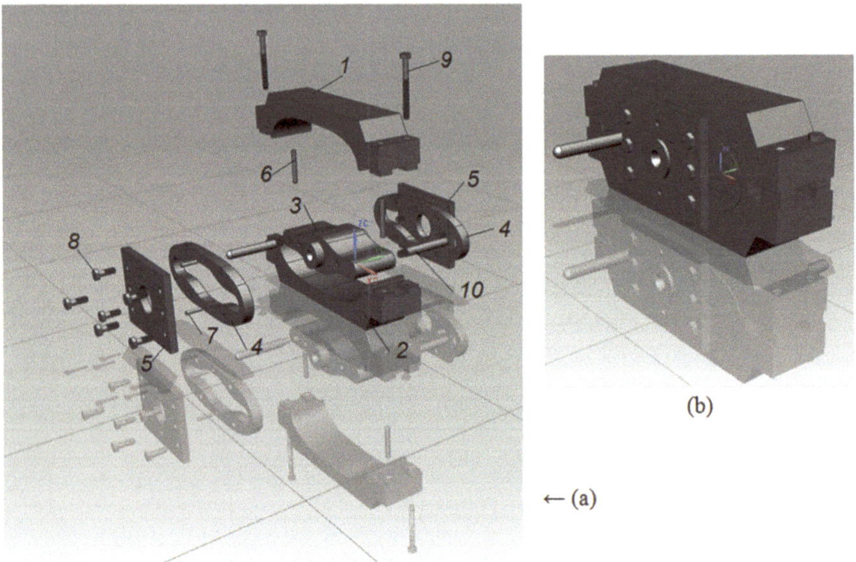

Fig. 4.10 Disassembled (**a**) and assembled (**b**) mold for curing the polymeric composite shell

optimization. The closed-mold design is presented in Fig. 4.9. It consists of two detachable half-molds, the mandrel, two caps and bolts, which serve to mold, assemble and pre-stress the prepreg. The convexity of the inner surface of the shell is ensured by the local layering of the additional snippets of reinforcing fabric between the layers of wound unidirectional tape.

The mandrel with wound prepreg is inserted into the mold (see Fig. 4.10), which is placed in a vacuum bag after assembling. The mold moves in the autoclave, where it is cured according to the predetermined temperature/pressure cure cycle. This cycle includes two temperature ramps with a heating rate of 2 °C/min, two dwell sections (isothermal holds) at 80 °C for 45 min and 150 °C for 2 h, and post-curing at 180 °C for 1 h to provide a better mechanical stiffness of the resin. After mold

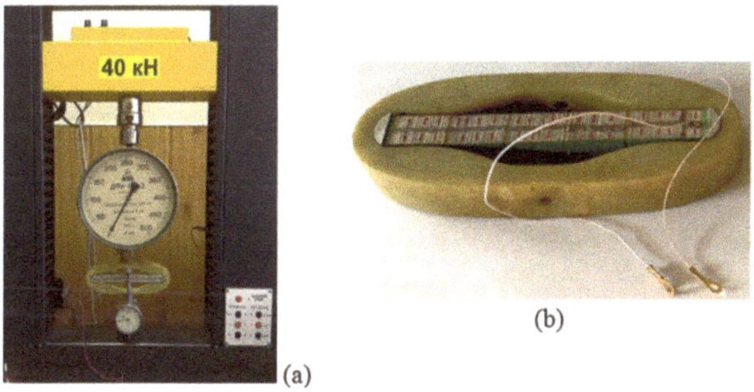

Fig. 4.11 Assembling of the actuator (**a**) and outline of the assembled actuator (**b**)

cooling and removing from the mandrel, the cured shell is subjected to milling of its lateral surfaces to achieve the required sizes and mutual parallelism of the lateral planes.

The piezoelectric stack consists of 240 piezoelectric layers 25 × 20 mm^2, each with 0.5 mm thickness, which are connected in parallel. The piezoelectric ceramics have properties similar to PZT-5H. In order to assemble the actuator and to provide pre-stressing of the PZT stack, the prepared shell was contracted along its shorter semi-axis by using the testing machine, and then the stack was inserted between two aluminum contact lugs (see Fig. 4.11). After unloading the shell, the stack becomes compressed under a force of about 2 kN. This pre-stressing removes the tensile strain, which is unacceptable for the brittle piezoelectric ceramics.

The completed actuators were then subjected to the static and dynamic testing.

4.4 Actuator Static Tests

In the static tests, the behavior of the actuator under external force action was investigated, in particular its stiffness and influence of the external force on the stroke. Dynamic tests aimed to determine the frequency properties of the actuator.

Since the stiffness of the stack does not affect the shell deformation under the action of an external compressive load on the shell along its minor axis, it was determined on the testing machine with a linear increase in the compressive load. The experimental results presented in Fig. 4.12 show that observable dependence "force—displacement" is linear in a wide range of external loads.

On the contrary, an external tensile force should transmit to the stack, causing its contraction due to compressive force, which acts along the PZT stack axis. This contraction depends on the mechanical stiffness of the piezoelectric and interlayer bonding material. These testing results are presented in Fig. 4.13.

To measure the tensile force, a force-measuring device of the testing machine was used, whereas the compressive force was measured using the small strain gauges, which were inserted between the PZT stack and aluminum contact lugs. The curve of linear dependence in Fig. 4.13 starts from the value of force ~2,300 N, which corresponds to the pre-stressing force.

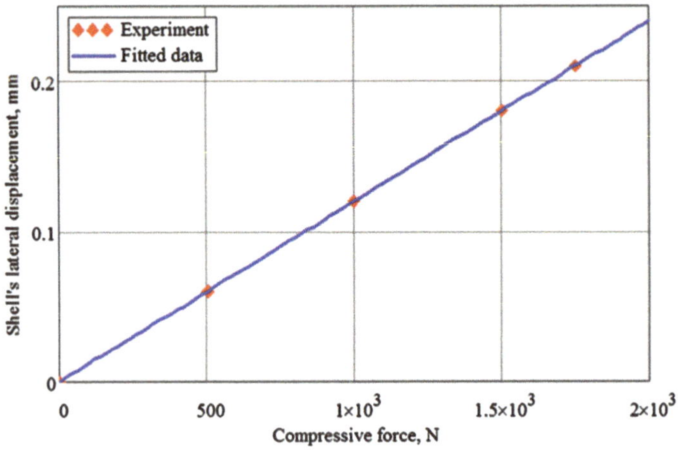

Fig. 4.12 Dependence of actuator's shell lateral displacement on external compressive loading

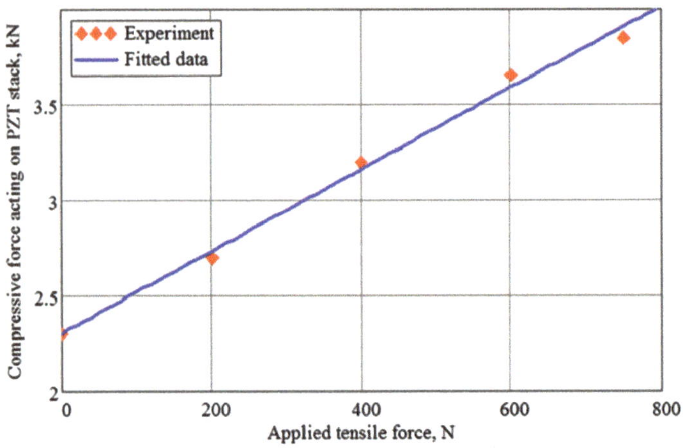

Fig. 4.13 Dependence of compressive force, acting on the PZT stack, on external tensile loading

The stroke dependence of the actuator on the driving voltage has been studied at the different values of external tensile force, which was applied to the actuating surface of the actuator by the cross-head of the testing machine. The tensile force was measured by the force-measuring device of the testing machine with a stepwise increase in the driving potential. Figure 4.14 demonstrates such a dependence, which is obtained for the initial value of external tensile force 600 N. It can be seen that this force grows with an increase of the driving potential, and reaches ~850 N at a maximum permissible voltage of 500 V.

No hysteretic phenomena have been detected at the static testing in the ranges of driving potential (from +450 to −150 V) and loads (from +750 to −325 N).

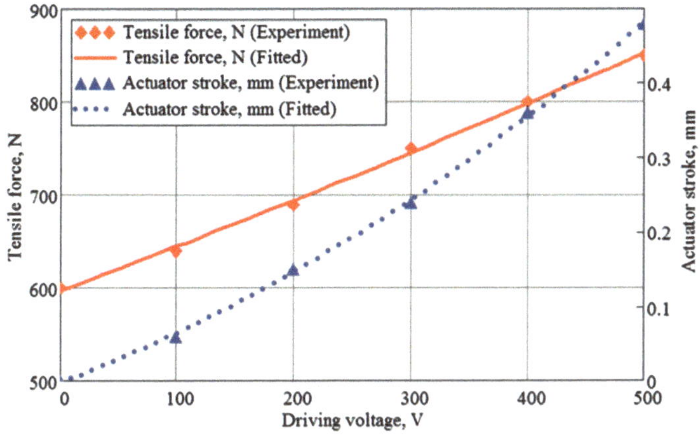

Fig. 4.14 Dependence of the actuator's stroke on driving voltage at an action external tensile force of 600 N

The established actuator parameters were acceptable for the customer–aircraft designer, but the most important data were obtained in the dynamic virtual and experimental tests.

4.5 Numerical and Experimental Tests of the Actuator's Dynamic Properties

When modeling the actuator's dynamic properties, we varied the driving frequency and counteracting external force in the ranges mentioned above, whereas amplitude of the driving voltage was always supplied as sinusoid wave with the frequencies of 2, 5, 10, and 20 Hz and amplitude 300 V with offset 150 V due to requirements of the PZT material. It is important to note that at the numerical FE simulation, we did not take into account an internal resistance of an amplifier and also loss factors of polymeric composite, piezoelectric, bonding material, and aluminum alloy.

The typical time histories of the free actuator stroke together with PZT stack displacement are demonstrated in Fig. 4.15.

An influence of applied harmonic load on the actuator stroke, which is presented in Fig. 4.15, demonstrates that the amplitude of sinusoid wave changing force gradually increases and stabilizes after two periods of oscillations. This loading reduces the stroke of the actuator by approximately 20%. The significant fall of the operating stroke, resulting from the action of external force, is observed in both Figs. 4.15 and 4.16.

This result, which was subsequently disproved experimentally, can be explained by the assumption of electric power supply with an infinite performance, as well as the fact that the first natural frequency of the actuator according to its operating

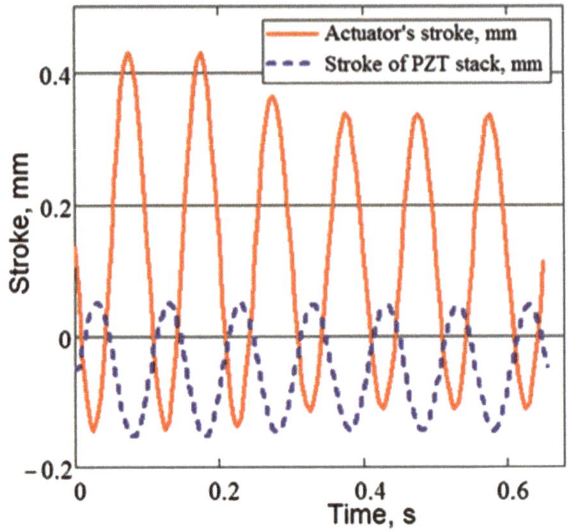

Fig. 4.15 Time histories of the actuator's parts displacements at a driving frequency of 10 Hz

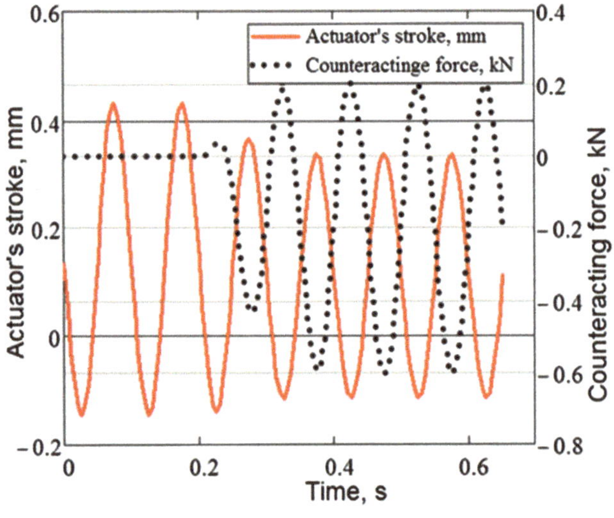

Fig. 4.16 Time histories of actuator's stroke and counteracting elastic force, applied to the actuator

movement was very high and equal to 680 Hz. In the framework of these assumptions, the operating actuator's stroke depended on the external counteracting force only. This dependence is shown in Fig. 4.17, and corresponds to the case of an ideal power supply and lack of mechanical energy loss that debilitates dynamic stiffness of PZT stack.

The time dependence of the mechanical power that the actuator generates is shown in Fig. 4.18. This curve corresponds to the case depicted in Fig. 4.16.

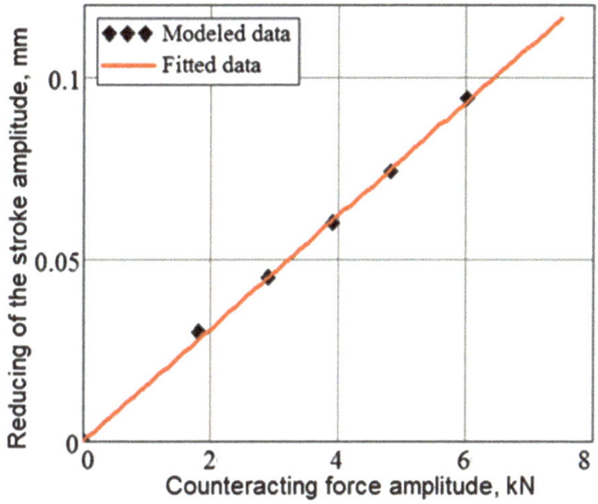

Fig. 4.17 Actuator's stroke dependence on counteracting loading force (FE simulation results)

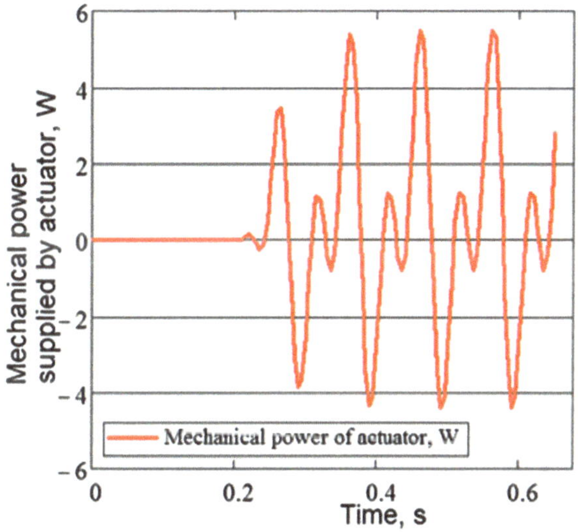

Fig. 4.18 Output mechanical power of actuator (FE simulation results)

Periodical change of the power sign is due to elastic resistance of the force, which acts against the actuator's stroke. According to the elastic nature of this force, for one-quarter of the period, the actuator overcomes the external force and for the remaining three-quarters of the period, the actuator moves along the force direction. The peak value of the output actuator's power depends linearly both on the acting force and driving frequency (see Fig. 4.19).

Time dependence of electric power, consumed by the actuator, is similar to the same value for mechanical power, which is presented in Fig. 4.18, but the amount of peak electric power is significantly larger. Moreover, the magnitude of this power is

Fig. 4.19 Frequency and external force dependencies on peak output mechanical power of actuator (FE simulation results)

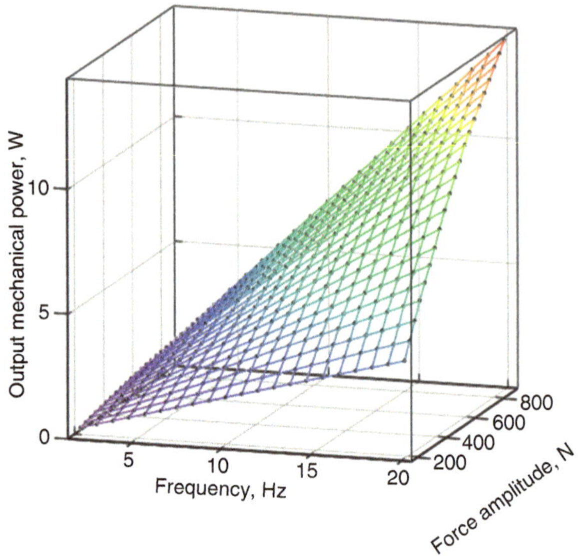

essentially independent of the load of the actuator. These results can be understood, taking into account the capacitive reactance of the PZT stack, which has electrical capacitance ~12 μF. Even a rough estimation for the reactive part of the electric power W^{react} and current I^{react}, consumed by this actuator at the limit frequency of 20 Hz (see Fig. 4.19) gives:

$$W^{\text{react}} \approx U^2\omega \cdot C = 450^2 \cdot 2\pi \cdot 20 \cdot 12 \cdot 10^{-6} \approx 305\text{W}$$

and

$$I^{\text{react}} \approx U\omega \cdot C = 450 \cdot 2\pi \cdot 20 \cdot 12 \cdot 10^{-6} \approx 0.7\text{A},$$

where ω is the angular frequency, and C is the electric capacitance of the PZT actuator. The active part of the power, which is consumed by the actuator for the operating displacement, is very small compared to the reactive electric power (Fig. 4.20). This estimation of required electric power and peak current imposes very high demands on the piezoelectric drivers.

These results have been verified during the experimental investigation, which was performed using the experimental set-up presented in Fig. 4.21.

Some dependencies of the actuator's stroke amplitude on the loading force and driving frequency demonstrated the significant influence of the amplifier parameter on the actuator performance. For example, even at the frequency of 2 Hz, the peak current is approximately 70 mA, which is close to the permissible limit for PA94 piezodrivers. This current is insufficient to supply the normal operation of the actuator. The actuator's required parameters have been reached using a PI E-617 High-Power amplifier, which provides an output electric power up to 280 W and peak current up

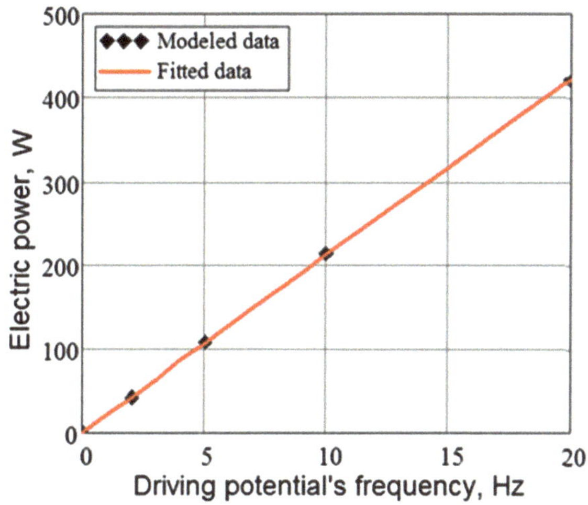

Fig. 4.20 Dependence of the peak electric power, which is consumed by actuator, on the driving frequency (FE simulation results)

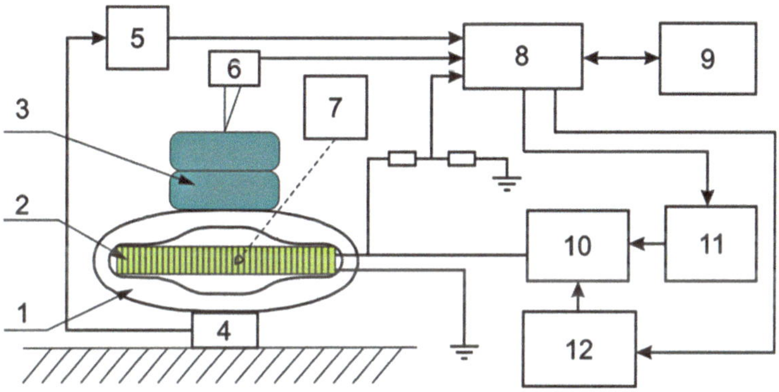

Fig. 4.21 Experimental set-up used to study the dynamic properties of actuator: 1—polymeric composite shell; 2—PZT stack; 3—damped mechanical load; 4—force-measuring device; 5—amplifier; 6—optical sensor; 7—temperature sensor; 8—ADC/DAC; 9—PC; 10—high voltage amplifier; 11—sinusoid wave generator; 12—electric power supply

to 2 A. By using one pair of these piezodrivers, the actuator's stroke changes from 10% to 25% compared to the values obtained by the numerical simulations.

4.6 Conclusion

To provide an actively controlled flap for a helicopter rotor blade, an optimization method for the amplified flextensional piezoelectric actuator has been developed. This kind of actuator has important advantages over the others, because it provides the greatest stroke amplification and has no joints with the friction surface. The proposed method is based on the representation of the shape of polymeric composite shell by the Cassini oval and the rational Bezier curves, whose parameters (coordinates and weights of the control points) are the design variables for the optimization algorithm. The optimization process is controlled by the Genetic Algorithm of Toolbox MATLAB, which uses the finite element model of the device, calculates the objective function (the stroke against given external load), and modifies the design variables to achieve the optimal solution. By using the finite element analysis of the developed structure, we established the best stiffness and stroke against the external forces. The finite element dynamic analysis and experimental study of the prepared actuator reveal the most important causes that reduce the performance of the actuator, e.g., the decrease of the multilayered PZT stack's stiffness, caused by the adhesive layers between PZT plates, and lack of peak output current and power, generated by the electronic driving system. This peak power of the driving system should be very high because of the great electric capacitance of the actuator's piezoelectric stack. We have also shown that by providing the required parameters of the driving electronics, the characteristics of the actuator should change little in the frequency range up to 20 Hz and more.

Chapter 5
Defects in Rod Constructions

5.1 Diagnosis of Defects and Monitoring of Rod Construction

The well-known classification of damage includes five levels [49, 146]. The first level consists of detection of defects in a construction without specifying their parameters. The second level determines the location of defects in construction elements. The third level assesses the degree of danger that the damage expresses . The fourth level is the most complex level, as the residual life of the construction as a whole is predicted. At the fifth level, the monitoring of construction defects is performed during loading construction over time.

Most investigations on the identification of defects are devoted to solving problems at the first and second levels [9, 11, 23, 31, 49, 70, 104, 105, 107]. Most of the studies were of the vibration parameters of rods and beams with open cracks. In addition, some papers consider finite-element models of elastic rods with cracks, the faces of which interact [23, 49, 104, 105, 107].

In a review of works on the identification of defects in constructions of various types [9], the results of solving problems at the five levels were analyzed. The different algorithms estimate the influence of the defect's parameters on the changes in the natural frequencies of oscillations, in the modal characteristics of the oscillation forms, and the change in the curvature of the forms of oscillation modes. The algorithms are based on dynamic calculations of matrices of compliance and damping of damaged structures [146]. The difference between the algorithms consists of the various expressions for the target functions and optimization schemes. In some algorithms for solving problems of damage identification, based on changes in the shapes of different modes, the modal assurance criterion (MAC) is used [9, 23, 146].

The known diagnostic signs of identification of defects in rod and beam constructions can be categorized into two groups, characterized both by the base on

S. N. Shevtsov et al., *Piezoelectric Actuators and Generators for Energy Harvesting*,
Innovation and Discovery in Russian Science and Engineering,
https://doi.org/10.1007/978-3-319-75629-5_5

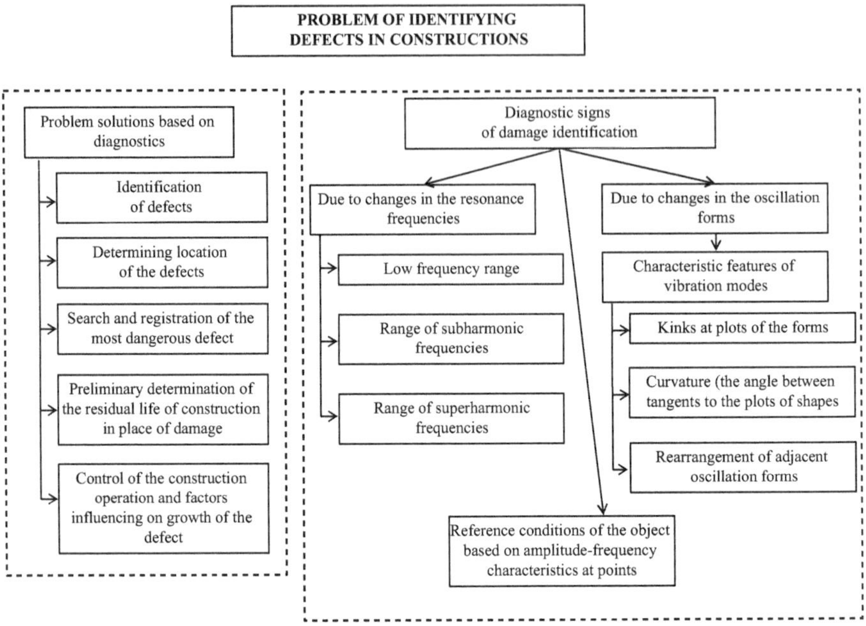

Fig. 5.1 Classification of problems and diagnostic signs in the identification of construction defects

which they are formulated and by the goals achieved in identifying defects [10, 16, 18, 40, 87, 153]. The combined classification of the most known diagnostic features is shown in Fig. 5.1. The above identification methods are also included in this classification scheme.

Consideration of various diagnostic signs of construction defects (included in Fig. 5.1) shows that the diagnostics of constructions can be made using the indicators of diagnostic reliability [10, 12, 67]. In particular, the problem of assessing the effectiveness of various diagnostic signs, which characterize the condition of constructions, has been previously discussed in [17]. This study was based on an analysis of known reliability indicators of control results. In this paper, it was shown that one way to improve the reliability of control results was to use complex methods based on the various diagnostic signs. The same conclusion was made in [67].

5.2 Reconstruction of Defect Parameters Based on Beam Models

Reconstruction, as well as the parameters of defects in rod constructions using the initial data on the oscillatory process, relates to the theory of elasticity and has great practical importance [175, 176].

It is well-known that, in a construction with defects, the parameters of oscillations change [175]. The oscillation parameters are viewed as the result of the action on the construction of the applied load in the assumption that the mechanical construction model is an elastic body. The damping of the oscillations is not taken into account. The forced oscillations are investigated by modal analysis and the natural frequencies with forms of oscillations are determined.

The result of identifying the damage parameters is the description of defect characteristics, namely the defect's location, structural sizes, and type.

By solving the inverse problems of identifying the defect's parameters, many changes can exist in the parameters of the oscillatory process of construction. Therefore, more than one defect is identified. Due to these circumstances, this type of problem is incorrect [174].

During operation, various types of defects arise in constructions of different complexities. The cracks resulting from fatigue or corrosion are the most common types of defects [101].

In practice, to identify defects in construction elements, the visual method is used [52, 125] (i.e., the defects are detected by sight). The shortcomings of such a method are that the defect's parameters need to have reached significant dimensions, have an open shape, and be visible on the external surface of the construction. In this case, a local change in the shape of the construction is possible when the limiting stress-strain state occurs at the stress concentration site. However, if the defect is located in an area out of sight, it will not be detected using the visual method.

To identify defects in construction, special diagnostic methods are used, e.g., the vibration method of defectoscopy [125, 127].

In a number of works [66, 109, 124, 134, 168–170], modeling of a construction defect in the form of a crack or notch is performed based on a local decrease in the stiffness of the cross-section of the element investigated.

In fracture mechanics [132], a crack modeling is performed for a certain configuration at specified loading parameters, and the following problems are considered: the stress-strain state near the crack, the degree of stress concentration at its tip, and the subsequent growth of the crack in the construction element is predicted [43, 85, 86, 159]. The application of the fracture mechanics method results in the evaluation of the durability of the construction [33]. In particular, the durability of the construction can be determined by calculating the growth rate of a crack, taking into account the intensity of the dynamic stresses at its tip [131]. Investigation of the first sign of the defect (crack) is conducted using the methods of the mechanics of deformable solids and mathematical modeling (i.e., the finite-element method) [19].

By modeling a defect in the form of a crack, the interaction of its faces can be taken into account, which leads to the complication of mathematical modeling of the oscillatory process and, as a result, does not allow for simple engineering solutions to the problems of fracture reconstruction [106, 109, 134, 172, 173]. In this case, there are changes in the rigidity of the damaged section and the nature of oscillations generated by the construction. Modeling a crack in the form of a notch that does not consider the opening–closing of the crack faces may limit the study of the physical processes occurring when the construction is dynamically loaded.

In [133], the authors considered the model of an equivalent elastic hinge element in describing the effect of crack parameters on the elastic compliance of the rod. In the calculations for the finite-element method, the technique for calculating the strength and rigidity of rod systems with the presence of cracks was considered [39].

In [65, 125], the authors describe the simulation of longitudinal vibrations in a rod with a crack defect. The crack had an open location. An approach to fracture modeling in the form of a longitudinal spring of a certain rigidity K was considered. The equivalent rigidity of an elastic element in the form of a longitudinal spring is assumed to be reverse-proportional to the defect size $d = 1/K$. The authors of [2, 27, 36] consider a one-dimensional system, described by a model of an equivalent rod system, divided into two parts and having a single defect. The connection between the rods (the defect model) is modeled as an elastic element (or spring) of a given stiffness K. A description of the dynamic characteristics for a system of equivalent rods is given, while the study of dynamic characteristics is performed for various natural frequencies of the system, in the presence of a defect.

In [121], the author described a rod with a constant cross-section free from fastenings at the ends and with a defect. The defect was modeled by an equivalent element in the form of an elastic spring element. In [115], a rod system is considered in the form of an equivalent model of a spring with a defect. The oscillations of the system are studied in the presence of external influences. The results of the changes in the two natural frequencies of the system depend on the location of the defect. The author concludes that the problem is incorrect and that if the system is symmetrical, then the presence of the defect at any of the symmetric points would lead to similar changes in the natural frequencies. Even if the system is unsymmetrical, a defect at different points can still generate similar changes in the natural frequencies.

In [116], the author considers rods with constant cross-sections and defects having different positions. The problem of the rod vibrations under different boundary conditions of fixation is considered. We consider cases in which the presence of information on the frequency change does not allow one to find the location of the defect.

In [117], the authors consider the problem of the vibration of a rod system, with the presence of a defect having point proof mass. The height and location of the proof mass, attached to the thin rod, are determined based on the defect's influence on natural frequencies. In [72, 114], the authors solve an inverse problem for determining the parameters of a defect (crack) in a rod. The location of the defect was uniquely determined using the asymptotic form of the spectrum. In [26], the authors investigated the asymptotic form of the spectrum for a homogeneous rod with a defect in the form of an equivalent spring and a rigidity K. The described rod structure had no fixation at the ends.

In [2, 68], the authors describe experiments to identify defects in a steel rod construction consisting of two parts. A procedure is given for the special case of identifying the defect's parameters that are dependent on natural frequencies.

The above publications only considered simple rod models. In the majority of the studies, the defect was modeled as an elastic element equivalent to a spring. Depending on the natural defect size and the spring rigidity, the behavior of individual vibration parameters of rod systems is not considered in the publications,. The authors confine themselves to the consideration of one parameter (the change in frequencies) depending on the spring stiffness, while not studying factors such as the shape of the oscillations and the variation of other parameters of the spring.

5.3 Reconstruction of Defects Based on Finite-Element Modeling

In [34–37, 51, 74, 92, 93, 119, 140, 152, 177, 178, 183], the authors analyzed the dependence of the oscillation parameters on the defect parameters based on finite-element modeling. An approach is considered by using the assumption that the stiffness matrix and mass matrix of the system have deviations to some values ($K + \delta K$, $M + \delta M$) and solutions of inverse problems are searched for. The changes in the various natural frequencies are calculated. It is assumed that the changes in frequencies are due to changes in the parameters of all or part of the elements of the finite-element model. Different methods of analysis determine the elements, whose changed parameters cause changes in a set of frequencies, while a statistical estimation of the solution is performed.

In [44, 45, 47, 65, 71, 127, 142, 151], the authors consider various ways of modeling the parameters of bending vibrations under impact loading of a beam construction with different fixings. At the same time, the authors of [28, 44, 121, 122, 181] modeled vibrations of a rod with a defect due to impact excitation. The defect was modeled as a simple equivalent spring element with a certain flexural rigidity K. The papers present various solutions for the problem of identifying the defect's parameters. The most developed method is described in [28]. The authors describe the subsequent stages of the solution of the simulation problem of construction vibrations and establish equations for determining the frequencies of a homogeneous beam with a defect in the form of an equivalent spring element with a flexural rotational rigidity K, located in a specific place R. Modeling of construction vibrations is performed at impact impulse loading and the analysis of the first six natural frequencies and their changes is carried out. In the results, the rigidity and location of the defect are defined.

In [113], the authors analyze a thin straight beam with a defect in the form of a spring having flexural rotational stiffness, and located in a certain site. The defect size depends on its rigidity and is investigated at comparatively small values of rigidity. Parameters of infinitesimal oscillations are analyzed without taking into account damping at a certain frequency. Transverse oscillations are described using the Euler-Bernoulli hypothesis. The authors note that the changes in the square of the frequency are proportional to the potential energy stored at the site of damage of the beam. They are also proportional to the square of the curvature of the shape of the vibration at the location of the damage when compared to a location without a defect.

In [116], the author considers a homogeneous freely-supported beam construction with a defect. The hypothesis is that the magnitude of the rigidity and the location of the defect in the beam are determined uniquely (with the exception of symmetry) by changes in the m-th and 2 m–th frequencies. By considering this beam structure, an alternative identification of the oscillation parameters with changes in the m-th and $(m + 1)$-th frequencies of the beam with sliding boundary conditions is given. The only conditions are that the oscillation form is defined by a separate sinusoid. At the same time, the vibration form includes sinusoid and cosine components in the common case. The proposed procedure can easily be generalized to fit the case of frequency changes in the presence of other boundary conditions. In [118], the authors use this approach to identify the parameters of damage in a beam construction with freely-supported ends in the presence of two defects.

In [128, 142], the authors showed that for the first mode of vibration in the case of a damaged beam, the oscillation curve increases in the region near the defect's location. In some forms of oscillation, the curvature does not change near the defect zone.

In [54, 55], the authors investigated the vibrations of a rod with a defect. In applying the Sturm theory, it was shown that the nodal points aspire to approach the site of the defect on the forms of the corresponding oscillations. The shape of the beam oscillation obeys the fourth-order equation, but not the simple second-order equation describing the rod. This approach is confirmed by the results of the [68] study.

In [91], the properties of fourth-order equations are considered. The results of the paper show that there are no simple generalizations of the Sturm theory to such equations. The points of the oscillation curve, called identical points, are found. It is shown that these points aspire to approach the defect site, but the identical points do not have a clear physical interpretation. In support of this conclusion, the authors of [55] give examples showing that the nodes do not always approach the defect site.

Different studies [29–31, 56, 78, 88, 126, 145, 147] consider finite-element modeling of rod construction and describe a crack in the rod under dynamic loading of the structure. This approach provides a sufficiently accurate and detailed study of the physical processes occurring near the defect location in the form of a crack. As an example of the simulation of fine-element modeling, the authors in [126] considered finite-element modeling simulation of a cantilever beam with a crack. This beam performed flexural vibrations under the action of a test harmonic force. A finite-element model of a cantilevered beam with a crack is shown in Fig. 5.2.

In modeling rod structures with defects, two models are generally used: piecewise linear and finite-element. The main differences in these models are as follows.

By using a piecewise linear model, only transverse deformations near the crack location that occur when bending vibrations of the rod structure, are taken into account. In contrast, the finite-element model has a more complex description structure and allows us to take into account not only longitudinal but also transverse shear deformations, as well as the effect of dry friction forces in the mutual slipping of the crack faces. By using the finite-element model, a more accurate study of the distribution of the stress-strain state in the area of the defect (crack) is provided [41, 42].

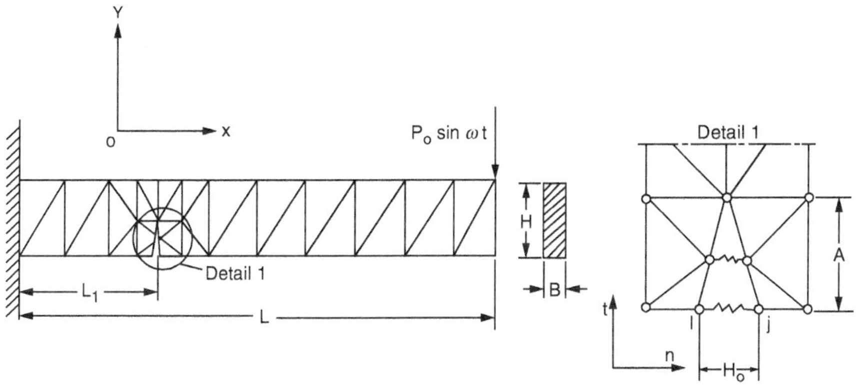

Fig. 5.2 Finite-element model of cantilever beam with crack

The application of the finite-element model of a crack allows us to describe its parameters quite accurately. This model is accompanied by a more complex mathematical description and application of specialized computer technology. Because of this, it is necessary to take into account the degree of complexity of the research for each specific case of the constructions with defects.

In research of the dynamic loading area of a construction, a refined finite-element model can be used for a detailed analysis of the distribution of stresses in the defect (crack) zone. Without the use of finite-element modeling, it is impossible to take into account the shear deformations and the forces of dry friction in the crack zone, whose influence on the dynamic response of the system, in particular, becomes especially noticeable in the analysis of flexural vibrations of short rods. The authors of [167] showed that the use of a piecewise linear model of a rod structure can be justified in cases where the length l is much larger than the cross-section size h ($l/h > 20$), while the effect of transverse shear deformations is very insignificant.

The use of mathematical modeling in identifying the parameters of defects in rod constructions allows us to identify new diagnostic features and therefore defects at various stages of their growth. The most commonly solved problem is the identification of the defect's parameters at its early stage of propagation. By investigating the adequacy of the two models, authors of [66, 172] showed that in a number of cases a piecewise linear model describing the construction with damage is sufficient. In this case, a quick speed of task solution is provided. By comparing two models (piecewise and finite-element) at the study of construction with damage the first model allows us to obtain sufficient for application results accompanied more quick solution of the problem.

The investigations carried out by the authors of [46, 166, 170] proved the efficiency of using a piecewise linear model in defect identification algorithms in rod constructions. For example, the authors of [31, 171–173] developed super-resonance and sub-resonance methods to search for defects in the form of cracks in rod elements of constructions, based on the use of a piecewise linear model.

Analysis of these studies shows that the best approximation is based on the finite-element approach [31]. However, there are difficulties in modeling and solving the inverse problem of finding the parameters of a defect due to the cumbersome numerical calculations, which require certain hardware and specific calculation software.

5.4 Goals of Following Study

The following two chapters present the results of the development of a technique for reconstructing defects in rod constructions, based on non-destructive testing methods, physical experimental modeling, and calculations using the chosen mathematical model. We shall evaluate the adequacy of the selected model to refine experimental research and to optimize the rod construction. Mathematical modeling includes the mathematical formulation of the problem, the development of methods for its solution, and the creation of software. Calculation of the characteristics of construction oscillations is carried out by solving the system of differential equations of the theory of elasticity together with the initial and boundary conditions. With this approach, we use direct numerical methods of the finite-element type. The implementation of this approach in the modeling of rod constructions of actual objects and the search for algorithms requires powerful computer resources and corresponding software, including service programs, database programs, specialized calculation programs, etc.

Moreover, the theory development and creation of an original information-measuring system for diagnostics and monitoring of defects in constructions is based on vibration parameters and the results of theoretical and experimental studies.

The specific issues presented in Chaps. 6 and 7 include the following:

(i) Methods for calculating the vibration parameters of undamaged and defective elastic rod structures.
(ii) Effects of the structural parameters of defects on the modal characteristics of models and information criteria that allow detection of defects based on the analysis of vibration parameters and the effectiveness of application of the criteria.
(iii) Algorithms for detecting defects in rod-shaped structures using vibration parameters.
(iv) Computer software with visualization of results, providing implementation of algorithms for detecting defects in rod constructions.;
(v) Methods and devices of the instrument-measuring complex for diagnosing defects in rod constructions.

Chapter 6
Identification of Defects in Cantilever Elastic Rod

6.1 Mathematical Formulation of the Problem of Defect Reconstruction in Cantilever

In a direct problem, steady-state vibrations of an elastic body are considered in a region with boundaries and they are described by the following boundary-value problem:

$$\sigma_{ij,j} = -\rho\omega^2 u_i, \sigma_{ij} = c_{ijkl}u_{k,l}, \ \ i = 1, 2, 3; \tag{6.1}$$

$$u_i|_{Su} = u_i^{(0)}, \ \ \sigma_{ij}n_j|_{St} = p_i, \ \ \sigma_{ij}n_j|_{S_\pm} = q_i, \tag{6.2}$$

where u_i are the searched components of displacement vector; $u_i^{(0)}, p_i, \ q_i$ are the known components of the displacement vector and surface loads; $\sigma_{ij}, \ c_{ijkl}$ are the components of stress tensor and elastic constants; ρ is the density; ω is the circular oscillation frequency; and $S_\pm$ is the internal surface of crack.

The considered physical model is shown in Fig. 6.1, for which the boundary is rigidly fixed. A force is applied at the point, varying according to harmonic law. In the boundary conditions (Eq. 6.2), the right-hand sides have the form:

$$\begin{aligned} &u_i^0 = 0, \ \ q_i = 0, \ \ p_2 = -Pe^{i\omega t}\delta(x_1 - l)\delta(x_2 - h)\delta(x_3 - 0.5a), \\ &p_1 = p_2 = 0. \end{aligned} \tag{6.3}$$

Identifying the parameters of defects (cracks, notches, inclusions, and cavities) requires determination of their configuration, so the surfaces are unknown, which relates the problems under consideration to the inverse geometric problems of the theory of elasticity. Further, we will assume that the crack faces do not interact and are free of stresses.

S. N. Shevtsov et al., *Piezoelectric Actuators and Generators for Energy Harvesting*, Innovation and Discovery in Russian Science and Engineering,
https://doi.org/10.1007/978-3-319-75629-5_6

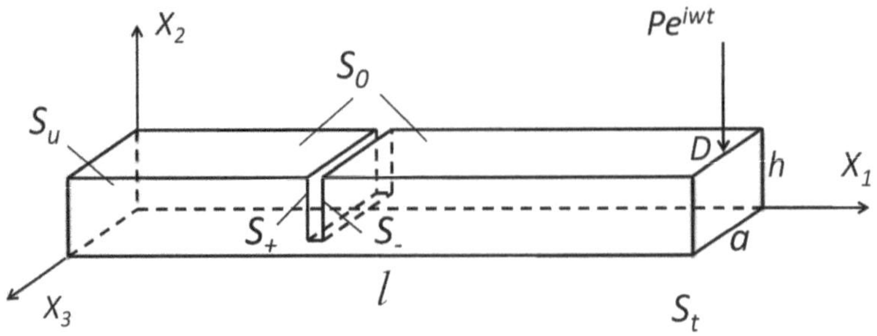

Fig. 6.1 Rod model with defect (in the form of notch)

To solve the inverse problems of surface reconstruction, some additional information is needed, which will serve a wave field of displacements $\bar{U} = (U_1, U_2, U_3)$, measured on a part S_0 of a free surface:

$$u_i|_{S_0} = U_i(x, \omega) \text{ at } x \in S_0 \text{ and } \omega \in [\omega_b, \omega_e], \tag{6.4}$$

or set of natural angular frequencies Ω:

$$\Omega = \{\omega_{r1}, \omega_{r2}, \ldots \omega_{rN}\}. \tag{6.5}$$

The values required in Eqs. (6.4) and (6.5) are easily measured in a full-scale experiment; in this case, a set of displacement amplitudes Ψ can be measured in Eq. (6.4) at steady-state oscillations in certain set of points x_k (positional scanning) and at angular frequency set ω_m (frequency scanning):

$$\Psi = \{U_i(x_K, \omega_m)|k = 1, 2 \ldots, K, \; m = 1, 2, \ldots, M\} \text{ at } x_k \in S_0 \text{ and } \omega_m \in [\omega_b, \omega_e]. \tag{6.6}$$

Ω or Ψ represents a set of input information for mathematical methods of defect reconstruction.

6.2 Finite Element Modeling of Cantilever with Defects and Analysis of Vibration Parameters

6.2.1 Full-Body Rod Model with Defect

Passing from the rigorous formulation of the problem in Eqs. (6.1) and (6.2) to the weaker formulation, using the FEM, we obtain the following matrix equation:

$$(-\omega^2[M] + [K])\{U_0\} = \{F\}, \tag{6.7}$$

where $[M]$ is the mass matrix; $[K]$ is the stiffness matrix; $\{U_0\}$ is the vector of nodal amplitudes of unknowns; and $\{F\}$ are the amplitude values of nodal influences.

The resonance natural frequencies and eigen-modes of oscillations are selected from the solution of the system:

$$([K] - \omega^2 M)\{U_0\} = \{0\}. \tag{6.8}$$

Due to the most frequent defects being in the form of edge cracks and notches on the surface of one of the construction faces, we shall further consider a rod with a one-sided defect. The scheme of cantilever with rectangular cross-section (where $L = 0.25$ m, $h = 0.008$ m, $a = 0.004$ m) is shown in Fig. 6.2.

The defect in the form of a notch was located in the place $\bar{L}_c$, where $\bar{L}_c = L_c/L$. The vertical axis of the notch is perpendicular to the main axis of the rod. The width of the defect was assumed equal to $d = 0.001$ m. The magnitude of the defect was varied within $t \in (0.0001\text{–}1)h$ and the relative size of the defect was: $\bar{t} = t/h$. This and further mechanical properties of the models were similar to steel St 10 with a Young's modulus $E = 2.1$ MPa and a density $\rho = 7{,}700$ kg/m^3.

FE modeling was carried out by using finite-element software ANSYS for a full-body 3D model, considered on the base of the 3D finite element Solid92. The final element Solid92 has a tetrahedral shape with 6° of freedom at each of the nodes (Fig. 6.3).

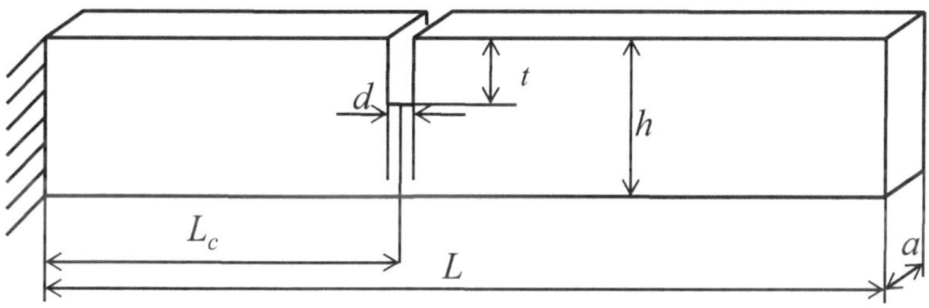

Fig. 6.2 Scheme of cantilever with defect

Fig. 6.3 Three-dimensional tetrahedral finite element Solid92 for constructing models in software ANSYS; circles denote the faces of the tetrahedron

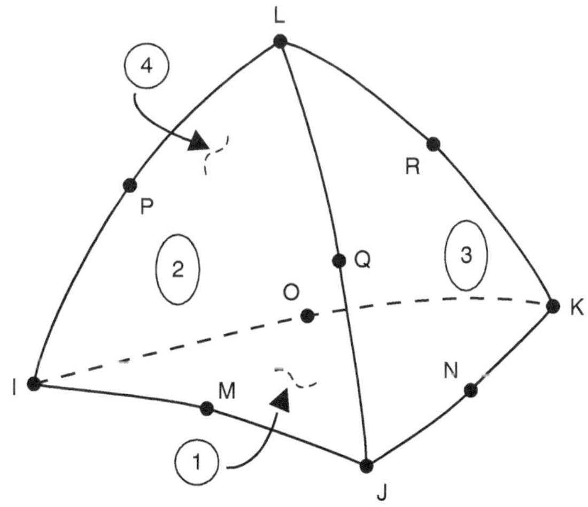

A preliminary analysis was made of the change in natural frequencies for various sizes of finite elements, both on the entire rod and near to the notch. The sizes of the finite elements were chosen in such a way that the error in determining the natural frequencies was minimal. It was chosen to divide the edges of the rod model into nodes along length by a factor of 1/30 of the rod length. The lateral edges along the height and width of the rod, as well as the edges of the faces, describing the defect, were divided into nodes by a factor of 1/3 of the corresponding length of the edge. The defect in the form of a notch presenting the full-scale model was modeled by a width of 1 mm perpendicular to the cross-section. The finite-element mesh had a double concentration near the defect. In this case, the total number of finite elements exceeded 5,000. A conditional partition into finite elements of the model is shown in Fig. 6.4. An example of a finite-element partition of an area with a defect in the form of a notch is shown in Fig. 6.5.

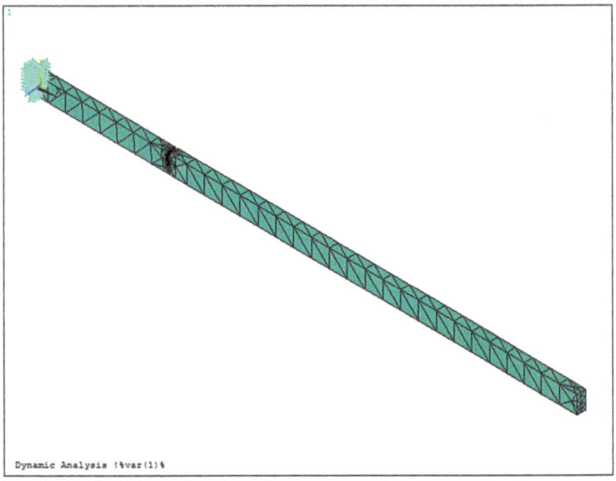

Fig. 6.4 Partition of rod model into finite elements in ANSYS

Fig. 6.5 Example of FE net near defect area (for defect size $\bar{t} = 0.7$)

In the results of the modal FE calculation of the vibrations of a rod with a defect in the form of a notch, natural frequencies and corresponding vibration modes were obtained. To obtain the amplitude-frequency characteristics at certain points of the vibrations of a rod with a defect, a spectral FE calculation was carried out.

6.2.2 *Analysis of Modal Parameters of Full-Body Model with Defect*

The research purposes consisted of obtaining the criteria for identification of defect parameters based on the analysis of the various modes of natural oscillations by applying a finite-element modal calculation of a cantilever full-body model of a rod with a single defect at different locations.

With the help of a finite-element software ANSYS, a full-body model of a rod without a defect is considered. The corresponding natural frequencies for the first 26 modes of natural oscillations were obtained. Tables 6.1, 6.2, 6.3, and 6.4, respectively, present the oscillation forms of the transverse vibrations of the rod model in the vertical OXY plane (Table 6.1), the transverse oscillations of the model in the horizontal OZX plane (Table 6.2), torsion vibration modes relative to the main axis of the rod OX (Table 6.3), as well as longitudinal oscillation modes (Table 6.4).

Analysis of the eigen-modes of the full-body model shows that the modes of natural oscillation in the OXY plane are 2, 4, 7, 10, 13, 17, 19, and 22. The modes of oscillation in the OZX plane are 1, 3, 5, 6, 9, 11, 14, 16, 18, 21, 24, and 26. Torsion vibration modes of oscillation are 8, 15, 20, and 25, and longitudinal modes are 12 and 23. Thus, when searching for the criterion of the presence of a defect, we can separately consider the forms of oscillation in different planes and axes of the model.

To analyze the sensitivity of the change in the first 26 eigen-modes of the cantilever oscillations on a single defect, we consider the dynamics of the change of the natural frequencies for the size of defect $\bar{t} = 0.9$ at different locations of the defect $\bar{L}_c = \{0.05; 0.15; 0.25; 0.35; 0.45; 0.55; 0.65; 0.75; 0.85; 0.95\}$. The relative changes in frequencies $\Delta\omega_p(\bar{t})$ were calculated from the formula:

$$\Delta\omega_p(\bar{t}) = \frac{\omega_p^i - \omega_p^o}{\omega_p^0} 100\%, \tag{6.9}$$

where ω_p^o, ω_p^i are the resonance frequencies from on-defect and defect models, respectively.

Graphic interpretations of the relative frequency change depending on the location of the defect for different transverse modes in the OXY (vertical), OZX (horizontal), torsion, and longitudinal oscillation modes, relative to the OX axis are shown in Figs. 6.6, 6.7, 6.8, and 6.9.

Analysis of data on the relative change in frequencies allows us to conclude that for most modes, the relative change in frequencies does not exceed 10%. Modes for

Table 6.1 Plane oscillation modes in OXY plane

Number of plane oscillation mode	Number of natural oscillation mode	Frequency, Hz	Number of plane oscillation mode	Number of natural oscillation mode	Frequency, Hz
1	2	108	2	4	673
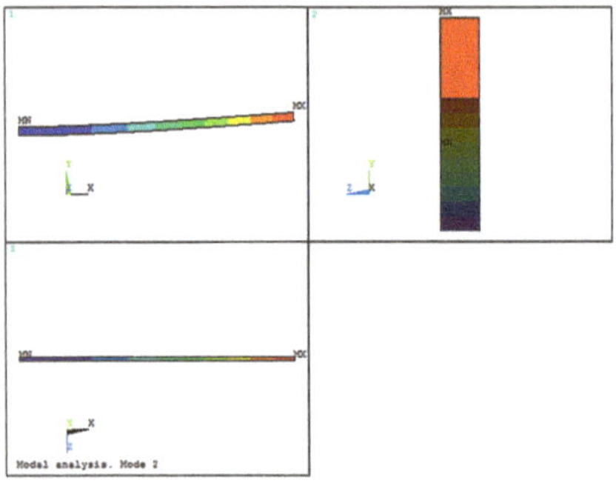			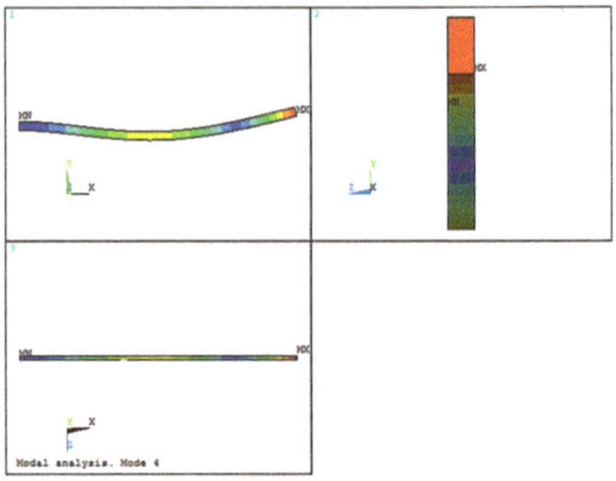		
3	7	1,872	4	10	3,603
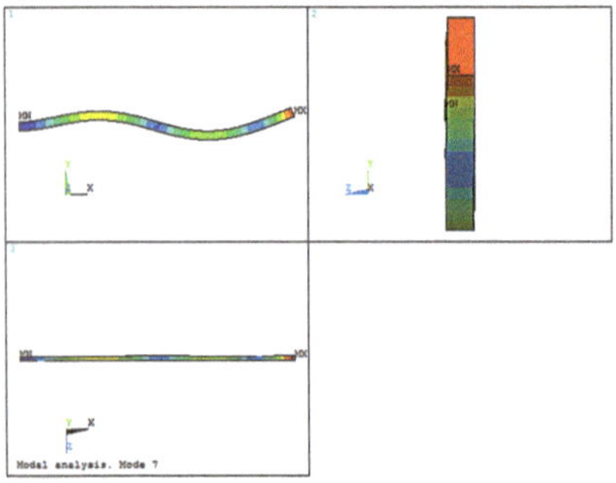			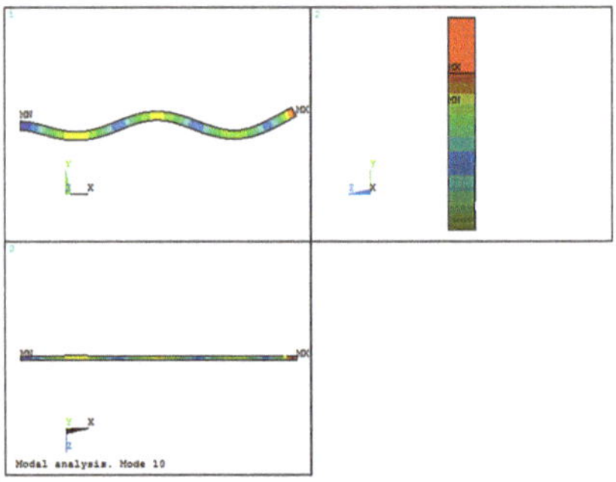		
5	13	5,919	6	17	8,701
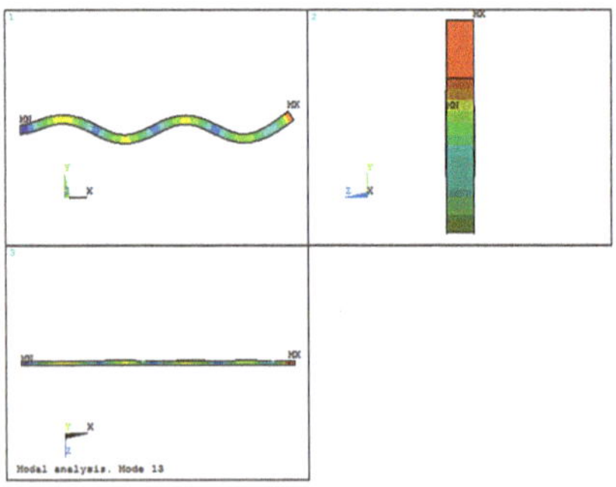			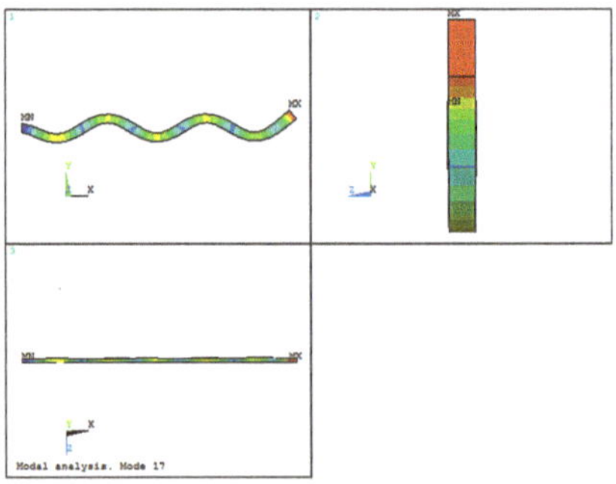		

(continued)

Table 6.1 (continued)

Number of plane oscillation mode	Number of natural oscillation mode	Frequency, Hz	Number of plane oscillation mode	Number of natural oscillation mode	Frequency, Hz
7	19	11,932	8	22	15,569
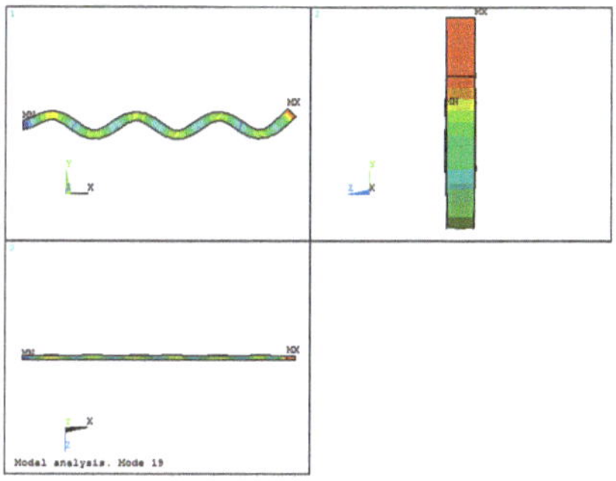			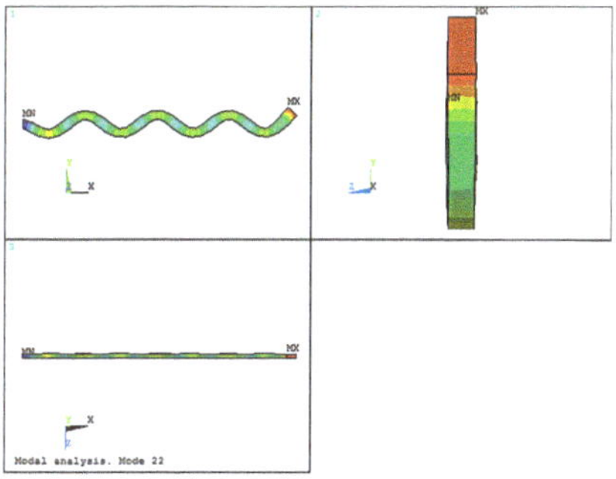		

oscillation forms in the OXY plane (2, 4, 10 19, 22), the OZX plane (1, 6, 25), for torsion vibration forms (8, 15), and for longitudinal vibration forms (12), have a high sensitivity to changes in frequency for different locations of the defect.

The problem of determining the sensitivity of the above modes was considered by identifying a defect in the model, taking into account that one of the most dangerous sections is the pinch region. When the defect (notch) is located near the pinch ($\bar{L}_c = 0.05$) for different values of the defect $\bar{t} = \{0.1; 0.3; 0.6; 0.9\}$, resonance frequencies $\omega_p(\bar{t})$ were obtained, as well as their relative values $\Delta\omega_p(\bar{t})$.

Figure 6.10 shows the plots of changing the first seven resonance frequencies $\omega_p(\bar{t})$ on the defect size $\bar{t}$ at the notch location $\bar{L}_c = 0.05$. Moreover, Fig. 6.11 presents plots of the relative change in frequencies on the defect size of the most sensitive vibration modes from the first 26 resonance modes of the cantilever.

The most sensitive modes of vibration to the defect in the cantilever for a given location of the notch are 1, 2, 4, 8, 12, and 22. The criterion for identifying the presence of a defect can be a sharp change in the first mode of vibration with a defect size of 0.6. Up to this defect size, the relative frequency change for the first mode of oscillation lies in small limits up to 2.5%.

Analysis of the problem solution showed the following: by searching for the defect test criterion, we separately considered 26 oscillation modes depending on the defect size and its location. Analysis of the data on the relative change in frequencies shows that the modes for the oscillation forms in the OXY plane (2, 4, 10 19, 22), in the OZX plane (1, 6, 25), for torsion (8, 15), and longitudinal oscillation (12) forms, have a frequency variation in the range 10–60% at different locations of the defect. At the same time, for 16 oscillation modes, the relative change in frequencies does not exceed 10%. When the defect (notch) is located near the pinch ($\bar{L}_c = 0.05$), it is revealed for different cut values that the most sensitive modes of vibration to the defect size for a given notch location are modes 1, 2, and 4.

Table 6.2 Plane oscillation modes in OZX plane

Number of plane oscillation mode	Number of natural oscillation mode	Frequency, Hz	Number of plane oscillation mode	Number of natural oscillation mode	Frequency, Hz
1	1	54	2	3	338
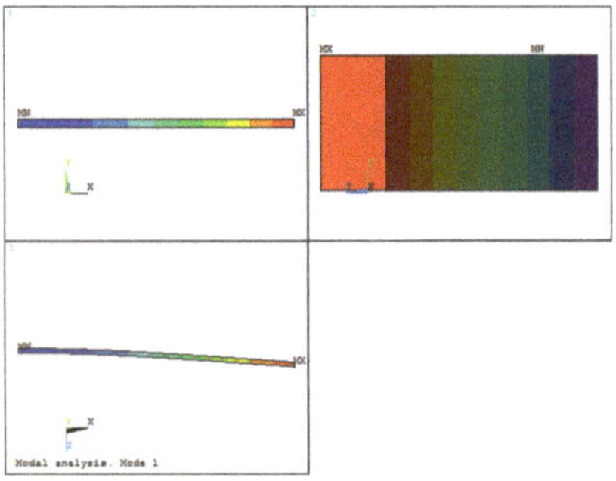			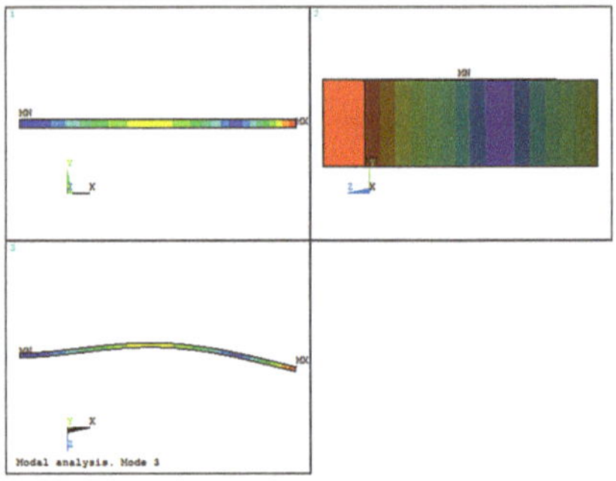		
3	5	946	4	6	1,849
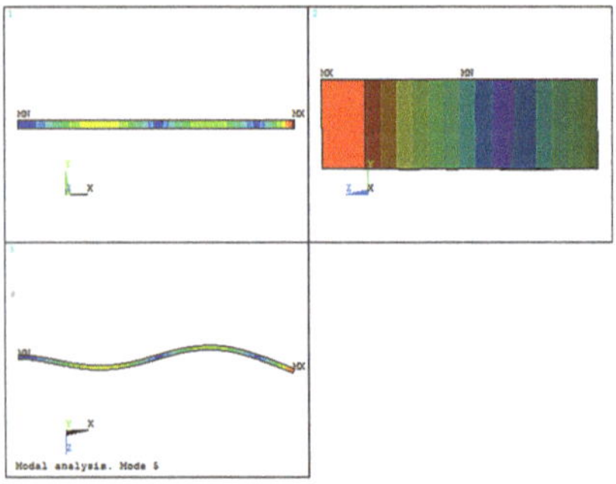			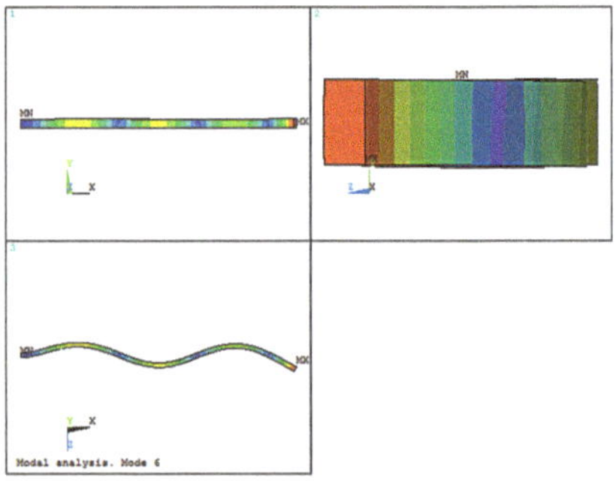		
5	9	3,047	6	11	4,534
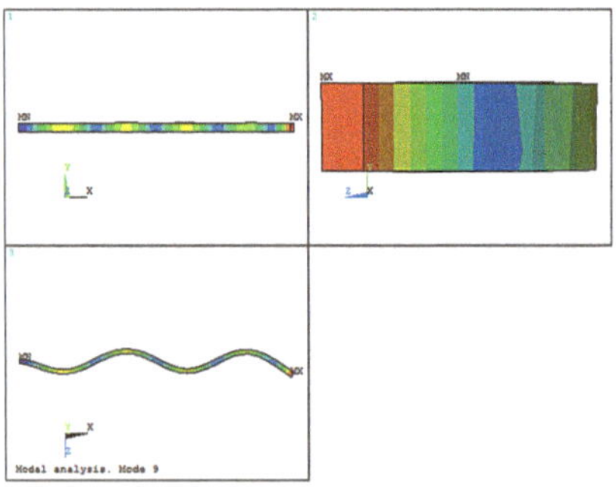			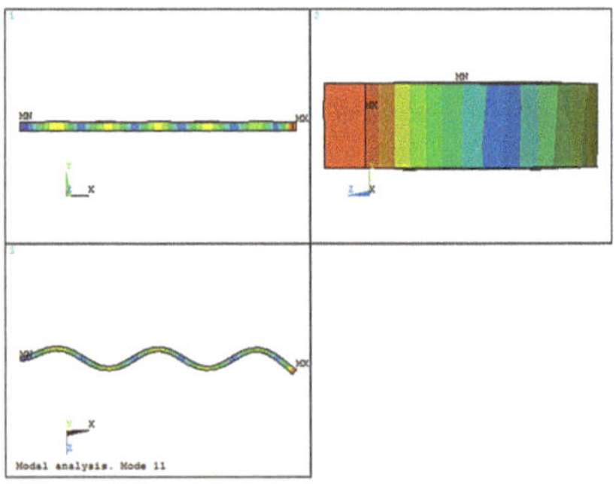		

(continued)

Table 6.2 (continued)

Number of plane oscillation mode	Number of natural oscillation mode	Frequency, Hz	Number of plane oscillation mode	Number of natural oscillation mode	Frequency, Hz
7	14	6,302	8	16	8,346
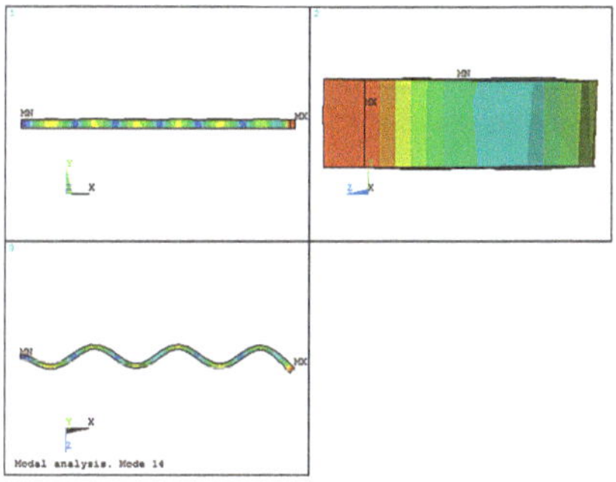			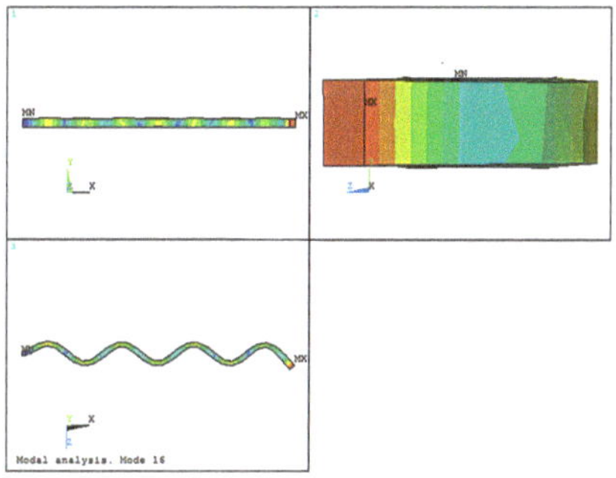		
9	18	10,654	10	21	13,220
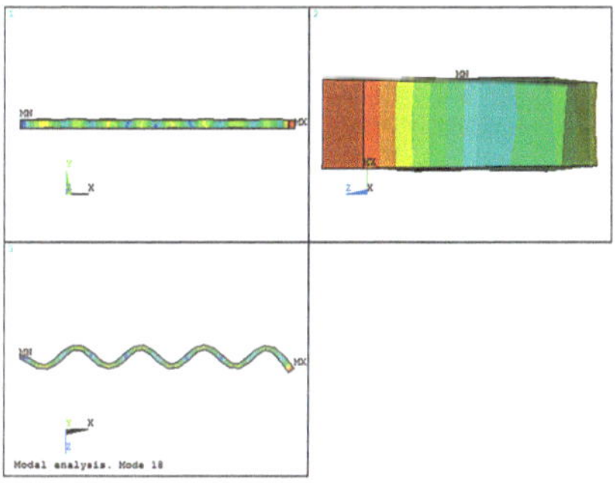			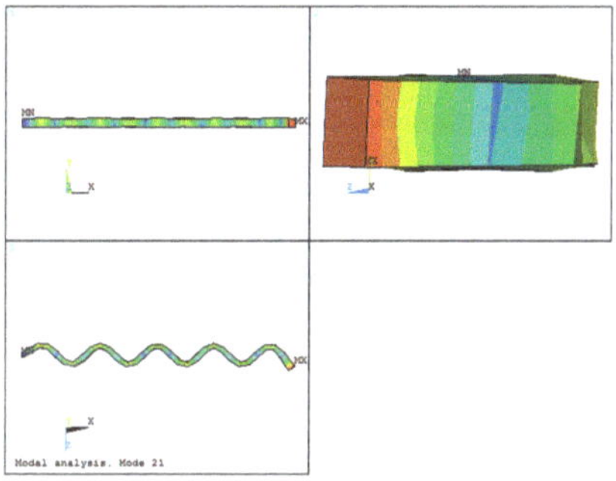		
11	24	16,033	11	26	19,080
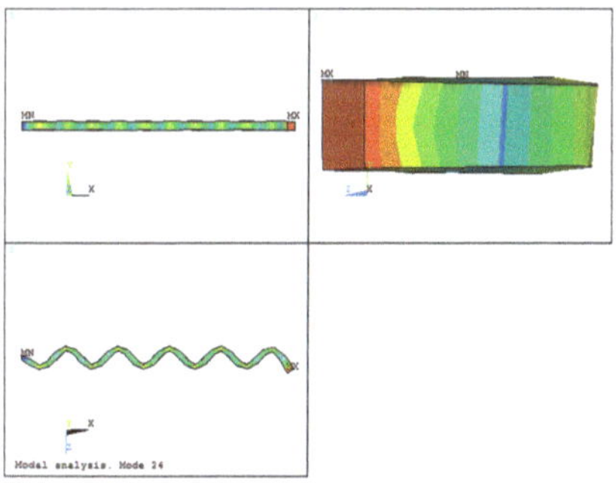			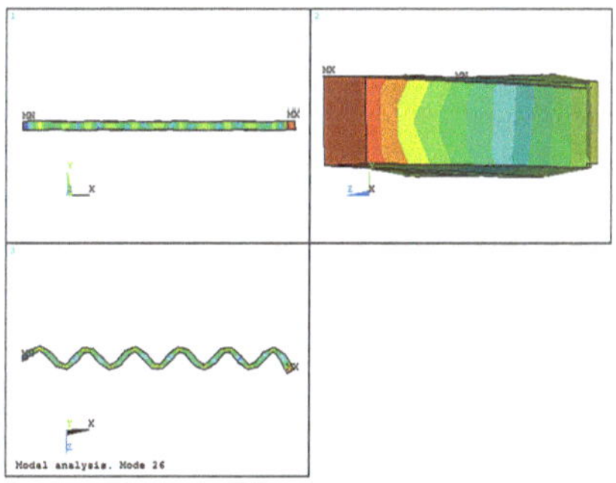		

Table 6.3 Torsion oscillation modes

Number of torsion oscillation mode	Number of natural oscillation mode	Frequency, Hz	Number of torsion oscillation mode	Number of natural oscillation mode	Frequency, Hz
1	8	2,410	2	15	7,232
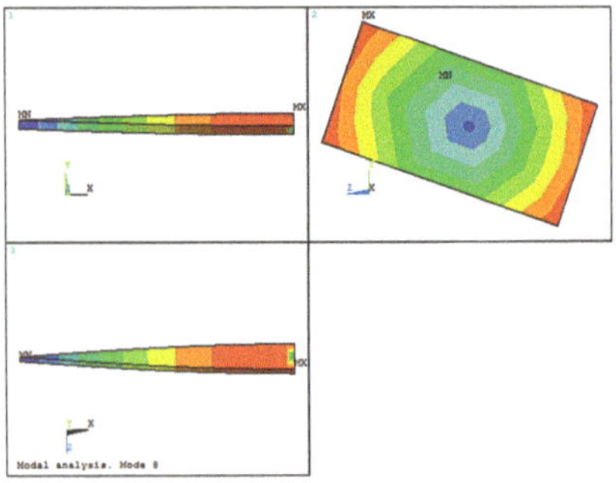			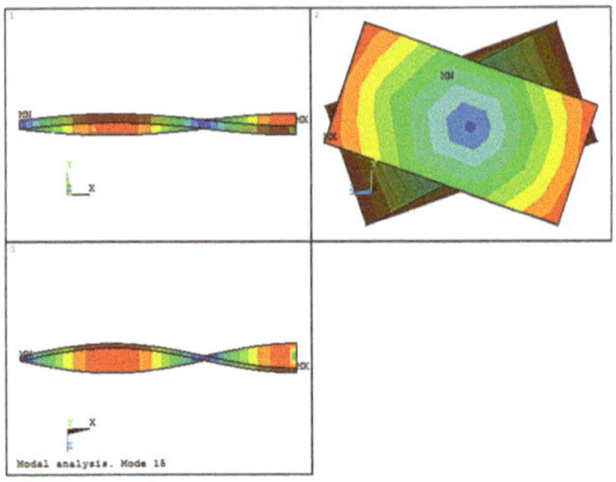		
3	20	12,058	4	25	16,892
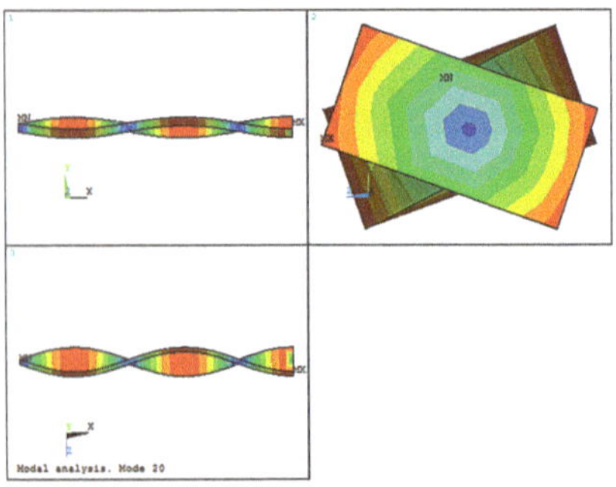			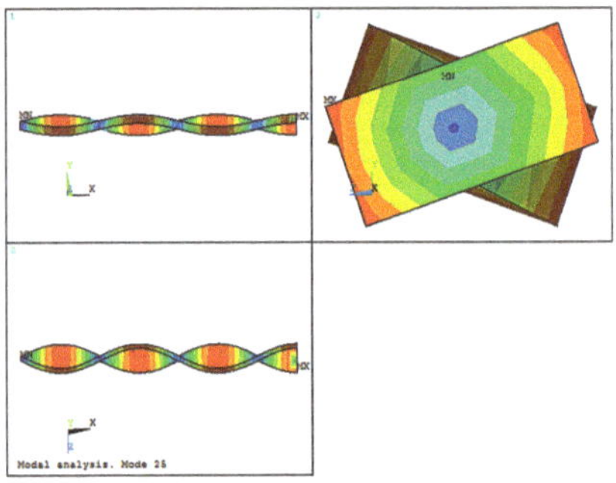		

6.2.3 *Comparison of Modal Parameters of Oscillations with Stress-Strain State of FE Cantilever Model with Various Notches*

By modeling a construction by using simplified models, the question arises on the dependency of various oscillation parameters on the notch shape in the cross-section of the rod structure. Here we consider defects, located at one site along the length of the rod, whose axis is perpendicular to the main axis of the rod and having an opening from one and two sides. The research defines oscillation parameters in dependence on the type of notch.

The considered cantilever with notches is present at the scheme shown in Fig. 6.12. The rod had dimensions: $L \times h \times a = 0.250 \times 0.008 \times 0.004\ \text{m}^3$. The width of the incisions was assumed equal to $b = 1$ mm and their disposition with different sizes on one or both sides (h_1, h_2) was at the same point with the coordinate

Table 6.4 Longitudinal oscillation modes

Number of longitudinal oscillation mode	Number of natural oscillation mode	Frequency, Hz	Number of longitudinal oscillation mode	Number of natural oscillation mode	Frequency, Hz
1	12	5,227	2	23	15,680
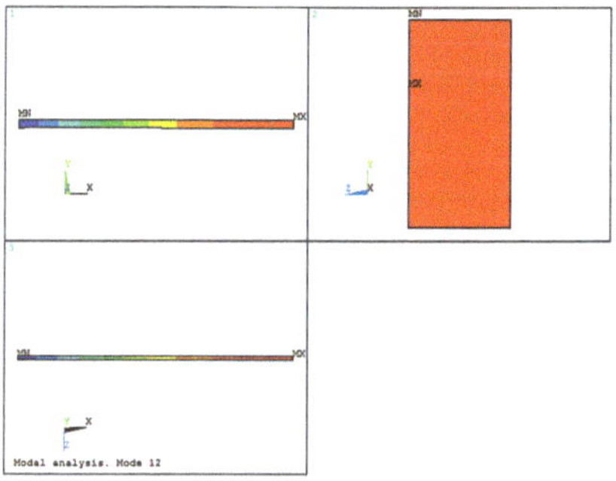			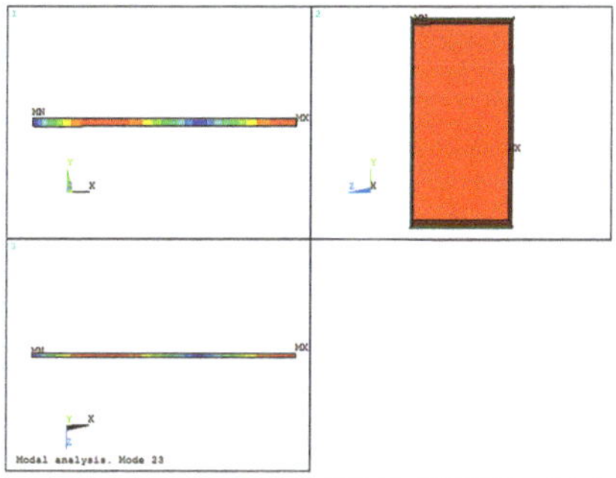		

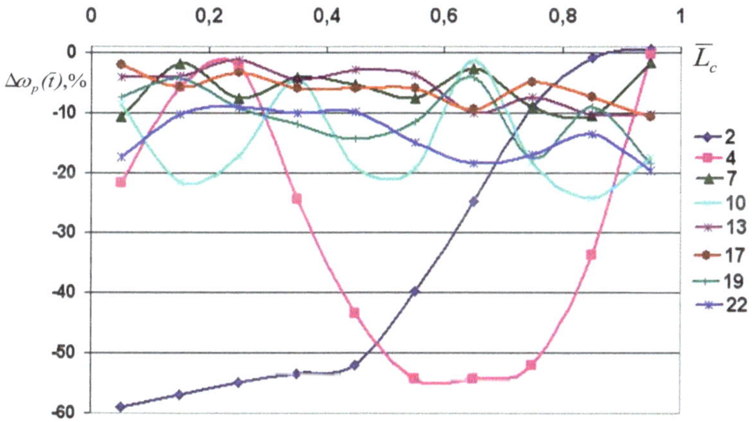

Fig. 6.6 Plots of relative frequency changes $\Delta\omega_p(\bar{t})$ for oscillation modes in the OXY plane for different defect location $\bar{L}_c$ and its relative magnitude $\bar{t} = 0.9$

on the horizontal axis $L_d = 0.0625$ m (relative size of the location of the notches to the length of the rod $\bar{L}_d = L_d/L = 0.25$). Next, we introduce a dimensionless coordinate $\bar{x} = x/L$.

The relative characteristics of the notch dimensions were considered, that is the sizes of the notches were normalized to the total height of the rod:

$$\bar{h}_1 = \frac{h_1}{h};\ \bar{h}_2 = \frac{h_2}{h};\ \bar{h} = \bar{h}_1 + \bar{h}_2.$$

The considered cases of notches are present in Table 6.5.

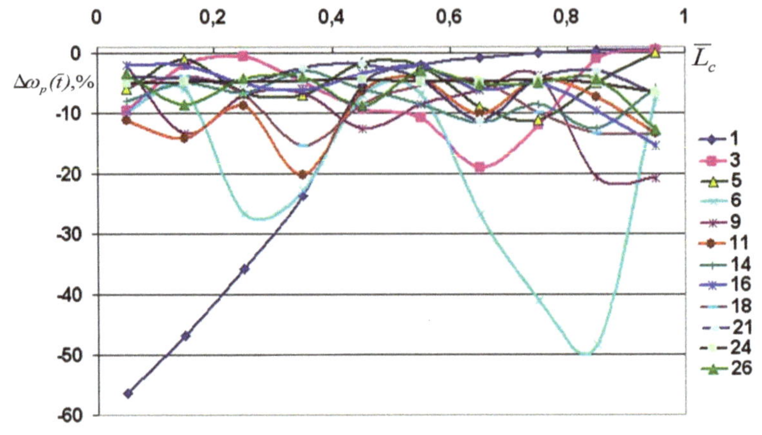

Fig. 6.7 Plots of relative frequency changes $\Delta\omega_p(\bar{t})$ for modes of oscillations in the OZX plane for different defect location $\bar{L}_c$ and its relative magnitude $\bar{t} = 0.9$

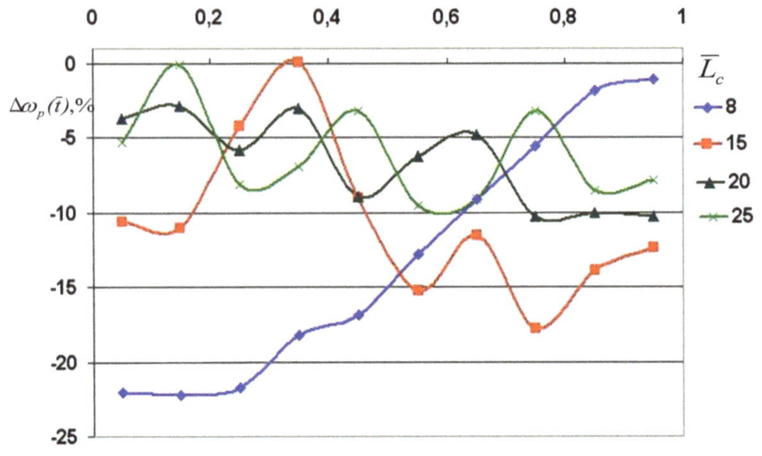

Fig. 6.8 Plots of relative frequency changes $\Delta\omega_p(\bar{t})$ for torsion vibration modes with different defect location $\bar{L}_c$ and its relative magnitude $\bar{t} = 0.9$

The simulation was carried out by using FE ANSYS software. We considered a full-body 3D model based on the application of the three-dimensional element Solid92. A preliminary analysis was made of the change in the natural frequencies for various finite element sizes, both over the entire rod and near the notches. In the case of finite element coding, the sizes of the finite elements were chosen in such a way that the error in determining the natural frequencies was minimal. The partition of the model into nodes along the length was made by a factor of 1/30 of the length of the rod. The height and width of the rod were divided into nodes by a factor of 1/3 of

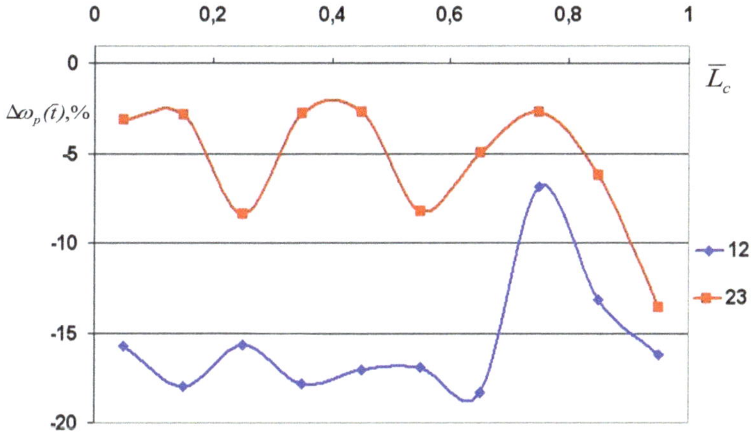

Fig. 6.9 Plots of relative frequency changes $\Delta\omega_p(\bar{t})$ for torsion vibration modes with different defect location $\bar{L}_c$ and its relative magnitude $\bar{t} = 0.9$

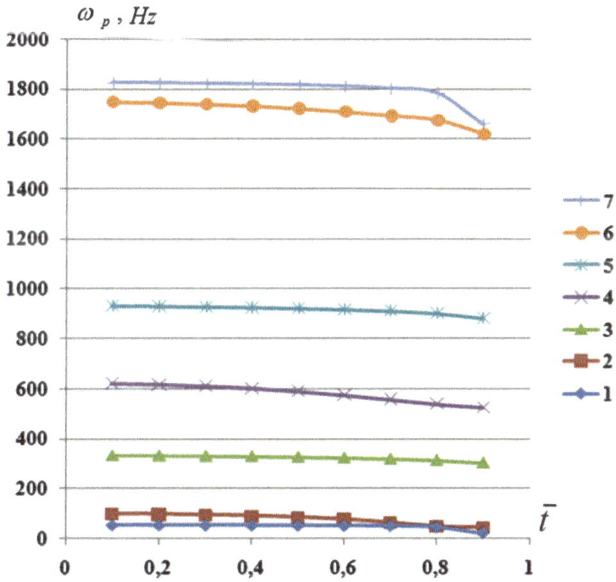

Fig. 6.10 Plots of first seven resonance frequencies $\omega_p(\bar{t})$ vs. relative of defect size $\bar{t}$ at its location $\bar{L}_c = 0.05$

the corresponding size of the face (Fig. 6.13a). At the place of modeling of notches, the finite element mesh was thickened (Fig. 6.13b).

Based on the FE modeling, modal calculation of the vibration parameters of the rod were performed. We considered forms of oscillations and natural frequencies of

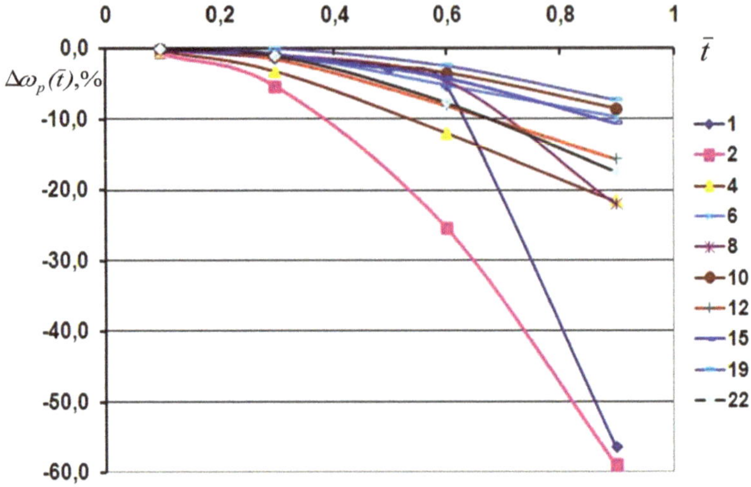

Fig. 6.11 Plots resonance frequencies for the most sensitive modes of vibration vs. defect size $\bar{t}$ at its location $\bar{L}_c = 0.05$

finite element models of cantilever with defects, located at the point $\bar{L}_d = 0.25$. Figure 6.14 presents curves of the transverse vibrations of the first mode in the plane of greatest stiffness (vertical plane) of the FE model near the defect. Tables 6.6 and 6.7 show the natural frequencies for the first mode of transverse oscillations of the model in the plane of greatest rigidity, the conditional amplitudes of oscillations, and the angle of the "breaking" oscillation shape at the location point of the notches with various variants $(\bar{h}_1, \bar{h}_2)$. These parameters were compared with the disposition of the defects symmetrically with respect to the horizontal axis of the rod. Analysis of the obtained resonance parameters shows that the natural frequencies, the oscillation amplitudes, and the angles of breaking the oscillation forms for the one-sided notch have the greatest deviation for notch sizes of $\bar{h} = 0.5$ $(\bar{h}_1 = 0, \bar{h}_2 = 0.5)$ and $\bar{h} = 0.7$ $(\bar{h}_1 = 0, \bar{h}_2 = 0.7)$. The relative deviations of the amplitudes of the oscillation shapes at the location point of the notches (Tables 6.6 and 6.7) in Figs. 6.14a, b are characterized by a corresponding "breaking" of the first oscillation form, and equal -6.5% and -17.3% for the notch sizes $\bar{h} = 0.5$ and 0.7, respectively. The relative deviations of the natural frequencies of the first oscillation mode (Tables 6.6 and 6.7) of the rod are equal to -3.27% and -7.44% for the notch sizes $\bar{h} = 0.5$ and 0.7, respectively. The relative deviation of the angles of the "breaking" the oscillation shape curve of the first mode is equal to -0.61% and -1.64% for the notch sizes $\bar{h} = 0.5$ and 0.7, respectively.

Tables 6.8 and 6.9 show the calculated resonances of the first ten modes of cantilever oscillations with different sizes of notches, and, respectively, their relative deviations from the case of symmetrical disposition of notches with respect to the horizontal axis of the rod $(\bar{h}_1 = 0.25, \bar{h}_2 = 0.25$ and $\bar{h}_1 = 0.35, \bar{h}_2 = 0.35)$. The

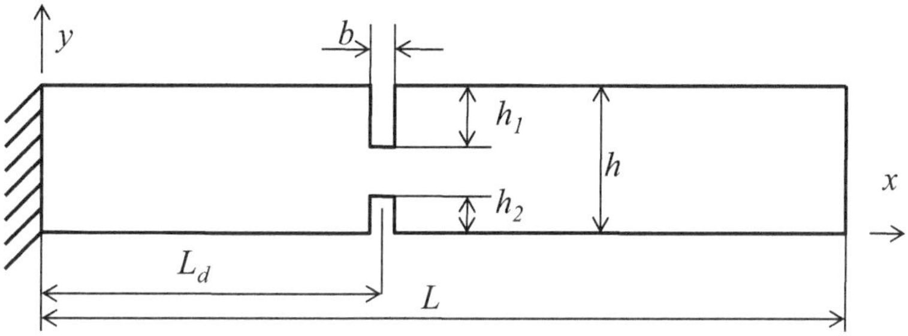

Fig. 6.12 Scheme of cantilever with notches

Table 6.5 Different variants of the relative values of the rod notches

Total length of notches	Variants of notches			
$\bar{h} = 0.50$	$\bar{h}_1 = 0.25$; $\bar{h}_2 = 0.25$	$\bar{h}_1 = 0.00$; $\bar{h}_2 = 0.50$	$\bar{h}_1 = 0.10$; $\bar{h}_2 = 0.40$	$\bar{h}_1 = 0.20$; $\bar{h}_2 = 0.30$
$\bar{h} = 0.70$	$\bar{h}_1 = 0.35$; $\bar{h}_2 = 0.35$	$\bar{h}_1 = 0.00$; $\bar{h}_2 = 0.70$	$\bar{h}_1 = 0.10$; $\bar{h}_2 = 0.60$	$\bar{h}_1 = 0.20$; $\bar{h}_2 = 0.50$

analysis shows that the greatest deviation of frequencies has a case of notches at $\bar{h}_1 = 0.0, \bar{h}_2 = 0.5$ and $\bar{h}_1 = 0.0, \bar{h}_2 = 0.7$, respectively, simulating a one-sided notch. In this case, the greatest deviation from the variant with the notches, located at the middle point of rod, has the entire frequency spectrum with a case simulating a one-sided notch.

For the case $\bar{h} = 0.3$, at the notch asymmetric shape and location $\bar{L}_d = 0.25$, the dependency of the first natural frequency on the notch width was analyzed. The results of calculating the first natural frequency with a change of the notch width in the range $b = 0$–0.001 m are given in Table 6.10.

Analysis of the change in the first natural frequency ω_1 depending on the notch size b at its location $L_d = 0.25$ shows that when the natural frequencies of the rod with the notch width $b = 0$ and $b = 0.001$ m are compared, the natural frequency decreases by 0.56%.

To compare the characteristics of the stress–strain state near defect for different defects at the location $\bar{L}_d = 0.25$, a calculation of stress–strain state was performed calculating the stress–strain state at static loading of the console sample by using ANSYS software. The load was modeled by a single force located at the free end of the rod, directed coaxially to the OY axis. The results of calculating the stress–strain state of the rod near the notch are shown in Fig. 6.15.

Analysis of the stress–strain state of the full-body rod with defects in the form of one-sided and two-sided notches shows that near the defect location within two sizes of the rod height, the stressed state differs from the stress state in the main beam. The area with the defect is small in comparison with the entire length of the rod.

(a)

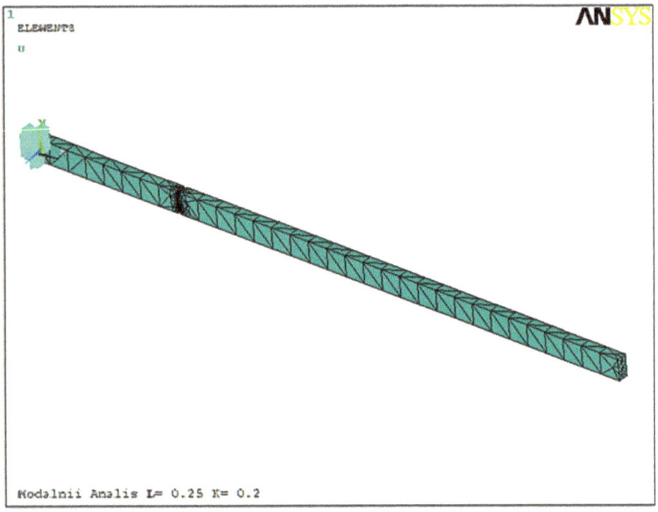

(b)

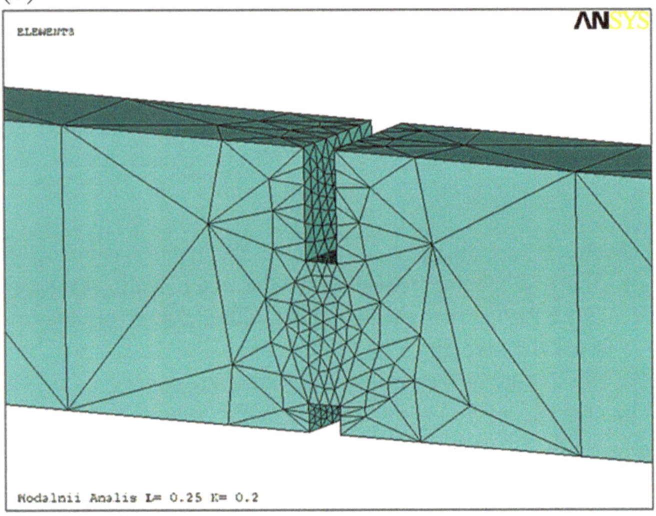

Fig. 6.13 Example of a finite-element partition of the rod model (**a**) and defect modeling corresponding to two notches, $\bar{h}_1 = 0.4$, $\bar{h}_2 = 0.1$(**b**)

Thus, by comparing the natural frequencies of the oscillations of the rod models with notches of different sizes, located at the place $\bar{L}_d = 0.25$, with the model of the rod having a symmetrical arrangement of the notches, the maximum deviations were obtained for all ten natural frequencies in the cases of simulating a one-sided notch as with $\bar{h} = 0.5$, and $\bar{h} = 0.7$. Comparative analysis of the shapes of the first oscillation mode showed that the greatest deviation of the oscillation amplitude for this case of

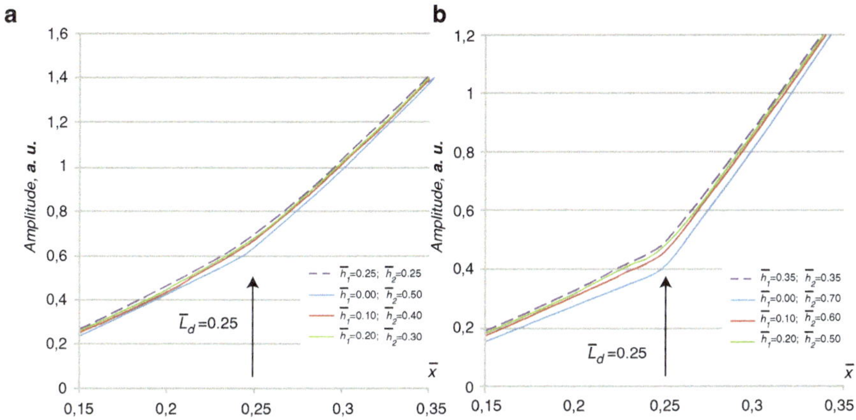

Fig. 6.14 Forms of the first mode of transverse oscillations in the plane of the greatest rigidity of model in the neighborhood of defect for its corresponding sizes: (**a**) $\bar{h} = 0.5$, (**b**) $\bar{h} = 0.7$

Table 6.6 Natural frequencies, conditional amplitudes and angles of the "breaking" first oscillation form at the point of notch location for total notch length $\bar{h} = 0.5$ and their relative deviations compared with the notch case ($\bar{h}_1 = 0.25$, $\bar{h}_2 = 0.25$)

Oscillation characteristics	Sizes of rod notches						
	$\bar{h}_1 = 0.25$ $\bar{h}_2 = 0.25$	$\bar{h}_1 = 0.00$ $\bar{h}_2 = 0.50$	Δ, %	$\bar{h}_1 = 0.10$ $\bar{h}_2 = 0.40$	Δ, %	$\bar{h}_1 = 0.20$ $\bar{h}_2 = 0.30$	Δ, %
Resonance frequency ω_1, Hz	100.9	97.6	−3.27	100.1	−0.80	100.7	−0.16
Conditional amplitude	0.680	0.636	−6.5	0.669	−1.7	0.678	−0.4
Angle of "breaking", degree	176.8	175.7	−0.6	176.5	−0.17	176.7	−0.05

Table 6.7 Natural frequencies, conditional amplitudes, and angles of the "breaking" first oscillation form at the point of notch location for total notch length $\bar{h} = 0.7$ and their relative deviations compared with the notch case ($\bar{h}_1 = 0.35$, $\bar{h}_2 = 0.35$)

Oscillation characteristics	Sizes of rod notches						
	$\bar{h}_1 = 0.35$ $\bar{h}_2 = 0.35$	$\bar{h}_1 = 0.00$ $\bar{h}_2 = 0.70$	Δ, %	$\bar{h}_1 = 0.10$ $\bar{h}_2 = 0.60$	Δ, %	$\bar{h}_1 = 0.20$ $\bar{h}_2 = 0.50$	Δ, %
Resonance frequency ω_1, Hz	84.81	78.50	−7.44	83.19	−1.92	84.83	0.02
Conditional amplitude	0.480	0.411	−14.3	0.462	−3.8	0.480	0.01
Angle of "breaking", degree	171.1	168.3	−1.64	170.3	−0.46	170.9	0.12

Table 6.8 Natural frequencies of first ten resonances for different cases of notch size $\bar{h} = 0.5$, located at $\bar{L}_d = 0.25$ for the FE rod model and their relative deviations compared with the defective rod, located symmetrically with respect to the horizontal axis of the rod ($\bar{h}_1 = 0.25$, $\bar{h}_2 = 0.25$)

	Sizes of rod notches						
Mode	$\bar{h}_1 = 0.25$ $\bar{h}_2 = 0.25$	$\bar{h}_1 = 0.00$ $\bar{h}_2 = 0.50$	Δ, %	$\bar{h}_1 = 0.10$ $\bar{h}_2 = 0.40$	Δ, %	$\bar{h}_1 = 0.20$ $\bar{h}_2 = 0.30$	Δ, %
i	ω_i, Hz	ω_i, Hz		ω_i, Hz		ω_i, Hz	
1	53.4	53.0	−0.64	53.3	−0.18	53.4	−0.02
2	100.9	97.6	−3.27	100.1	−0.80	100.7	−0.16
3	338.9	338.8	−0.04	338.9	−0.01	338.9	0.00
4	672.9	671.9	−0.14	672.7	−0.04	672.9	−0.01
5	941.0	936.5	−0.48	939.7	−0.14	940.9	−0.01
6	1,790.5	1,751.1	−2.20	1,780.4	−0.56	1,788.5	−0.11
7	1,841.3	1,831.5	−0.53	1,838.2	−0.17	1,840.7	−0.03
8	2,500.9	2,467.8	−1.32	2,490.1	−0.43	2,499.3	−0.06
9	3,062.4	3,057.9	−0.15	3,061.1	−0.04	3,061.8	−0.02
10	3,486.8	3,407.7	−2.27	3,464.1	−0.65	3,482.7	−0.12

the notch location took place for the one-sided notch and was $\Delta A = 6.5\%$ at $\bar{h} = 0.5$ and $\Delta A = 14.3\%$ at $\bar{h} = 0.7$; the greatest deviation of the angle of "breaking" of the oscillation form was $\Delta\alpha = -0.6\%$ at $\bar{h} = 0.5$: and $\Delta\alpha = -1.64\%$ at $\bar{h} = 0.7$.

By examining various notches, located at $\bar{L}_d = 0.25$, analysis shows that the nature of the stress state at the static loading and the parameters of the vibration modes differ only near the defect location. In this case, the resonance frequencies differ in small deviations from the case of the defect location at the middle of the rod, which can be used to apply a simplified beam model in calculating the oscillation parameters.

Consequently, it is assumed that the hypotheses, for example, of the Euler–Bernoulli theory cannot be satisfied in the neighborhood of the defect, and it is necessary to take into account this region by modeling the defect as a separate equivalent element. An example of such an element is an elastic spring with flexural rigidity.

6.3 Analysis of the Vibration Parameters of Cantilever with Defects Based on the Analytical Modeling

6.3.1 Identification of Cantilever Rod Defects Within the Euler–Bernoulli Model

Let us consider a cantilever rod, consisting of a homogeneous material with a defect in the form of a crack or notch, which opens during bending vibrations. Here, we present approaches to the identification process of defects in a full-body rod, based on the consideration of a simplified equivalent model, consisting of elementary links.

Table 6.9 Natural frequencies of the first ten resonances for different cases of notch size $\bar{h} = 0.7$, located at $\bar{L}_d = 0.25$ for the FE rod model and their relative deviations compared with the defective rod, located symmetrically with respect to the horizontal axis of the rod ($\bar{h}_1 = 0.35, \bar{h}_2 = 0.35$)

Mode	Sizes of rod notches						
	$\bar{h}_1 = 0.35$ $\bar{h}_2 = 0.35$	$\bar{h}_1 = 0.00$ $\bar{h}_2 = 0.70$	Δ, %	$\bar{h}_1 = 0.10$ $\bar{h}_2 = 0.60$	Δ, %	$\bar{h}_1 = 0.20$ $\bar{h}_2 = 0.50$	Δ, %
i	ω_i, Hz	ω_i, Hz		ω_i, Hz		ω_i, Hz	
1	52.5	51.8	−1.37	52.3	−0.53	52.5	−0.10
2	84.8	78.5	−7.44	83.2	−1.92	84.8	0.02
3	338.7	338.4	−0.10	338.6	−0.04	338.7	−0.01
4	668.9	667.2	−0.27	668.4	−0.08	668.9	−0.01
5	930.7	921.4	−1.00	927.1	−0.39	930.1	−0.07
6	1,637.0	1,573.4	−3.89	1,615.9	−1.29	1,634.1	−0.18
7	1,820.2	1,802.5	−0.97	1,813.4	−0.37	1,819.0	−0.07
8	2,425.9	2,340.0	−3.54	2,391.9	−1.40	2,418.0	−0.33
9	3,055.7	3,047.8	−0.26	3,052.7	−0.10	3,055.2	−0.02
10	3,287.9	3,136.0	−4.62	3,220.1	−2.06	3,268.9	−0.58

Table 6.10 First natural frequency and its change with notch size

No.	Notch width b, m	First frequency ω_1, Hz	Frequency change, $\Delta\omega_1$, %
1	0	105.46	0
2	0.0003	105.26	−0.19
3	0.0007	105.07	−0.37
4	0.001	104.87	−0.56

The considered system is shown in Fig. 6.16 and represents a cantilever rod having a rectangular cross-section with height h and width b. In the rod, there is a defect in the form of a notch, located at a distance L_c from the pinching. The form of the defect is considered open. The defect is located in the cross-section of the rod, perpendicular to its main axis.

A physical simplified equivalent model for calculation is the composite beam. Because of the limited size of the notch (crack), the model of a defective rod can be represented as a model having an elastic element in a certain damaged section with a flexural stiffness factor K_t (Fig. 6.17).

The rod element is loaded with a harmonic force $F_0 e^{i\omega t}$, where F_0 is the force amplitude, ω is the frequency of the oscillations, and t is the time. In this case, the system is schematically divided into three sections (Fig. 6.18): (i) pinching—elastic element; (ii) elastic element—force application point; (iii) force application point—free edge of the rod.

Let us consider the differential equation of forced oscillations in the framework of the Euler-Bernoulli model:

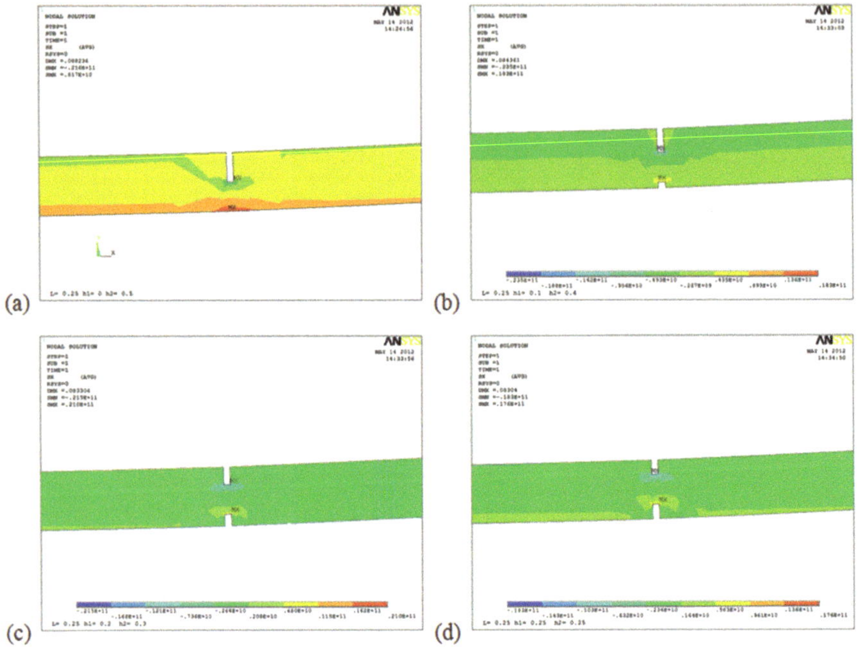

Fig. 6.15 Stress state of rod near notch at various sizes of this defect: (**a**) $\bar{h}_1 = 0.00$, $\bar{h}_2 = 0.50$; (**b**) $\bar{h}_1 = 0.10$, $\bar{h}_2 = 0.40$; (**c**) $\bar{h}_1 = 0.20$, $\bar{h}_2 = 0.30$; (**d**) $\bar{h}_1 = 0.25$, $\bar{h}_2 = 0.25$

$$\frac{\partial^2}{\partial x^2}\left[EJ(x)\frac{\partial^2 u_i}{\partial x}\right] - m(x)\frac{\partial^2 u_i}{\partial t^2} + F(t)\delta(x - L_F) + p(x,t) = 0, \qquad (6.10)$$

where $u_i(x, t)$, $i = 1, 2, 3$ are the displacements of points of the axis of the beam, where the subscript indicates the number of the section of the beam, as shown in Fig. 6.18; E is the elastic modulus; $J(x)$ is the moment of inertia of the section; $m(x)$ is the linear density; $F(t)\delta(x - L_F)$ is the force applied at a point; and L_F; $p(x, t)$ is the distributed load.

The boundary conditions for a composite construction have forms:

at $x = 0$:

$$u_1(0) = 0;$$
$$u_1'(0) = 0;$$

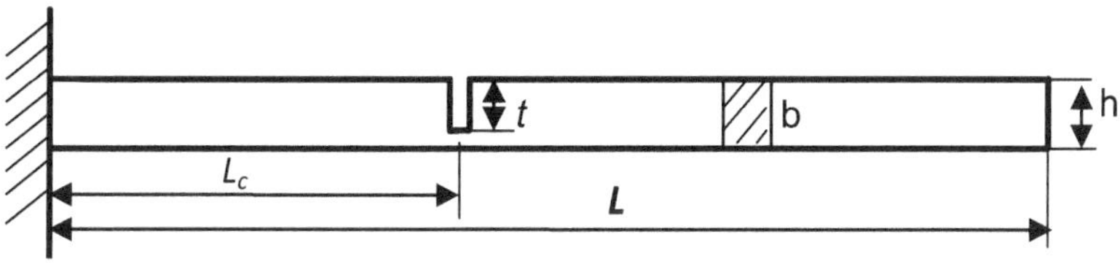

Fig. 6.16 Scheme of cantilever with defect (notch)

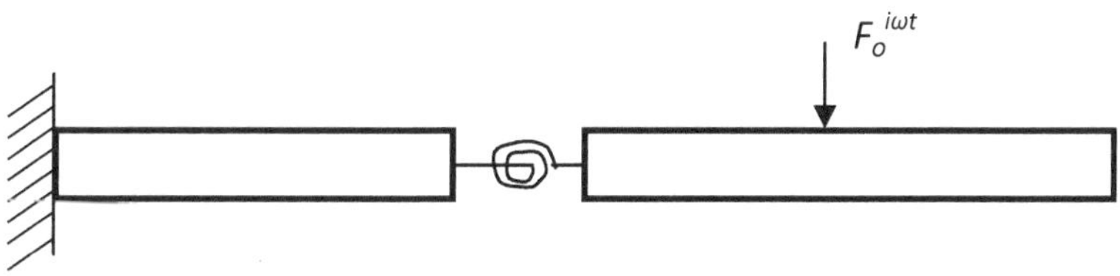

Fig. 6.17 Model of cantilever with elastic element

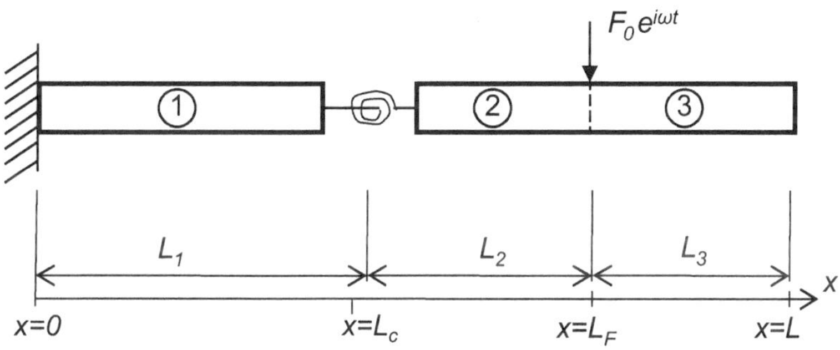

Fig. 6.18 Partition of rod system with elastic element into sections

at $x = L_c$:

$$
\begin{aligned}
u_1(L_c) &= u_2(L_c);\\
u_1''(L_c) &= u_2''(L_c);\\
u_1'''(L_c) &= u_2'''(L_c);\\
-EJu_1''(L_c) &= K_t\left[u_1'(L_c)u_2'(L_c)\right];
\end{aligned}
\tag{6.11}
$$

at $x = L_F$:

$$u_2(L_F) = u_3(L_F);$$
$$u_2'(L_F) = u_3'(L_F);$$
$$u_2''(L_F) = u_3''(L_F);$$
$$u_2'''(L_F) - u_3'''(L_F) = F_0/EJ;$$

at $x = L$:

$$u_3''(L) = 0;$$
$$u_3'''(L) = 0;$$

where K_t is the rigidity of elastic element.

We shall search all solutions in the following form:

$$u_i(x,t) = \sum_{k=1}^{\infty} u_i(x) F_0 e^{i\omega t}, \tag{6.12}$$

and have

$$\frac{d^4 u_i(x)}{dx^4} - \lambda_B^4 u_i(x) = 0, \tag{6.13}$$

where factor $\lambda_B^4 = \omega^2 \rho A l^4/(EJ)$; ω is the angular frequency of oscillations; ρ is the density of the material; $A = bh$ is the cross-sectional square of the rod; l is the length of the corresponding section of the rod; $J = \frac{bh^3}{12}$ is the moment of inertia of the section.

The solution of Eq. (6.13) in the absence of a distributed load and constants J and m, expressed in terms of Krylov functions $K_i(\lambda_B x)$, $i = 1, ..4$, is written as

$$u_i(x) = C_{i1} K_1(\lambda_B x) + C_{i2} K_2(\lambda_B x) + C_{i3} K_3(\lambda_B x) + C_{i4} K_4(\lambda_B x), \tag{6.14}$$

where C_{ij}, $i = 1, 2, 3$; $j = 1, 2, ..4$ are constants defined from the boundary conditions; $K_g(\lambda_B x)$, $g = 1, 2..4$ are the Krylov functions:

$$\begin{aligned} K_1(\lambda_B x) &= \frac{1}{2}(\mathrm{ch}(\lambda_B x) + \cos(\lambda_B x)); \\ K_2(\lambda_B x) &= \frac{1}{2}(\mathrm{sh}(\lambda_B x) + \sin(\lambda_B x)); \\ K_3(\lambda_B x) &= \frac{1}{2}(\mathrm{ch}(\lambda_B x) - \cos(\lambda_B x)); \\ K_4(\lambda_B x) &= \frac{1}{2}(\mathrm{sh}(\lambda_B x) - \sin(\lambda_B x)). \end{aligned} \tag{6.15}$$

Depending on the location L_C of the elastic element and application of force F_0, we can present the system of equations of the rod vibrations. The equations of motion for each of the rod sections are given as:

$$\begin{aligned} u_1(x) &= C_{11}K_1(\lambda_B x) + C_{12}K_2(\lambda_B x) + C_{13}K_3(\lambda_B x) + C_{14}K_4(\lambda_B x); \\ u_2(x) &= C_{21}K_1(\lambda_B x) + C_{22}K_2(\lambda_B x) + C_{23}K_3(\lambda_B x) + C_{24}K_4(\lambda_B x); \\ u_3(x) &= C_{31}K_1(\lambda_B x) + C_{32}K_2(\lambda_B x) + C_{33}K_3(\lambda_B x) + C_{34}K_4(\lambda_B x). \end{aligned} \quad (6.16)$$

Natural frequencies are found from equality to zero of the determinant of this equation set:

$$\Delta\left(\omega_i, K_t, \bar{L}_c\right) = 0, \quad i = 1, .., n. \quad (6.17)$$

To solve the inverse problem of reconstructing the spring stiffness and the length of the first section (for a given total length of the rod), a segment of the natural frequency spectrum is selected as additional information. Such data can be obtained experimentally as a result of processing the responses of a system with a defect in harmonic or non-stationary loading. The process of measuring the natural frequencies in the work was modeled through their calculation using ANSYS software for a rod element with a defect, as a three-dimensional body. Substituting Eq. (6.16) into Eq. (6.11), we obtain SLAE for the determination of arbitrary constants $C_{ij}(i = 1, 2, 3; j = 1, 2, ..4)$

The use of equations for natural frequencies ω_i in Eq. set (6.17), which demonstrate a significant dependence on the rigidity of the elastic element (i.e., defect size), overcomes two main problems in solving inverse problems: absence of the uniqueness of the solution and great sensitivity to the error in the input information.

6.3.2 Analysis of Sensitivity of Natural Frequencies to Size and Location of Defect in Analytical Modeling

Let us investigate the sensitivity of Eq. (6.17) to the value of the rigidity of the elastic element, and also to its location $\bar{L}_C$ for a cantilever-pinched rod of a rectangular cross-section with dimensions of $L \times h \times a = 0.25 \times 0.008 \times 0.004$ m^3 at location of force F_0 at the free edge of the rod (see Fig. 6.19).

Introducing a dimensionless coordinate $\bar{x} = x/L$, we consider Eq. set (6.17) with the following conditions:

$$\begin{cases} \Delta\left(\overline{\omega}_i, \bar{K}_t, \bar{L}_c\right) = 0, i = 1, 2, 3, 4; \\ \bar{K}_t \in [0.01, \ldots, 1]; \\ \bar{L}_c \in [0.01, \ldots, 0.99]. \end{cases} \quad (6.18)$$

In the equation, dimensionless parameters are considered, while the normalized value of the elastic element's rigidity was adopted as $K_t = 50000$ N $\cdot$ m/rad. The

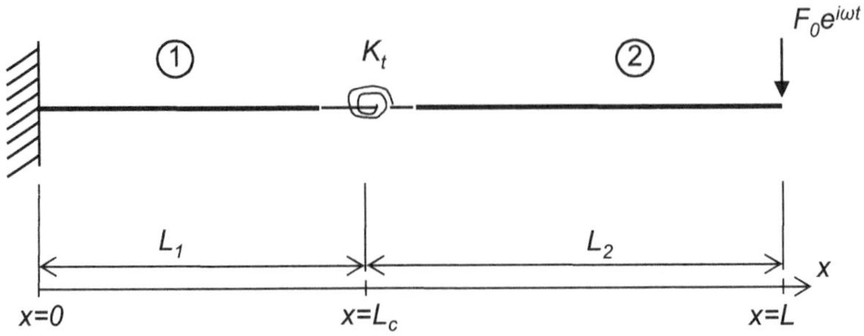

Fig. 6.19 Rod scheme with boundary conditions

oscillation frequencies were normalized to the frequency of the intact rod for each oscillation mode, respectively. We considered the first four natural oscillation frequencies. The oscillation frequencies for the intact model were calculated and amounted, respectively, to: $\omega_1^0 = 107.8$ Hz; $\omega_2^0 = 676.8$ Hz; $\omega_3^0 = 1{,}895$ Hz; $\omega_4^0 = 3{,}715$ Hz. Numerical determination of frequencies $\overline{\omega}_i$for given parameters $\bar{K}_t$ and $\bar{L}_c$ was carried out using Maple software. The results of the calculations are the surfaces of the dependencies of frequency $\overline{\omega}_i = \overline{\omega}_i(\bar{K}_t, \bar{L}_c)$ on the location $\bar{L}_c$ and rigidity $\bar{K}_t$ of the elastic element (Fig. 6.20).

Analysis of the dependencies shows that they have a complex spatial nature of frequency $\overline{\omega}_i$ variation, depending on the location $\bar{L}_c$ and rigidity $\bar{K}_t$ of the elastic element. Each plot has its own distinctive features.

For more detailed analysis of the dependencies, let us consider the cross-sections of these surfaces by planes: $K_t = \{1;\ 250;\ 1000;\ 5000;\ 50000\}$ N·m/rad.

For different stiffness values K_t of the elastic element, a graphic interpretation is obtained for the first four natural frequencies (Fig. 6.21). The results of the calculations are present in the plots of the dependences of the reduced natural frequencies of oscillations $\overline{\omega}_i = \omega_i^* \left(\bar{L}_c, K_{ti}\right)/\omega_i^0; i = 1, 2, 3, 4$, where ω_i^0 is the natural frequencies of the intact rod. As the plots (Fig. 6.21 a, b, c, d) show, the sensitivity of the frequency characteristics $\overline{\omega}_i$ of the rod for a different arrangement of the elastic element $\bar{L}_c$ is different and has a complex character of change.

Analysis of the frequency dependencies $\overline{\omega}_i$ for four natural frequencies (Fig. 6.21) showed the following: The investigated natural frequencies of oscillations depend in a complex way on the location $\bar{L}_c$ of the elastic element and its stiffness coefficient K_t. In particular, in some ranges of the position of the elastic element, the values of the natural frequencies diminish with decreasing the stiffness coefficient K_t of this element. In a set of other locations of the elastic element, the rigidity of the elastic element does not affect the values of the natural frequencies. Moreover, it follows from the nature of the frequency dependencies that the location $\bar{L}_c$ of the elastic element affects various natural frequencies in different ways. To quantify this feature, the plots of these dependencies were processed. The results of this treatment are given in Table 6.11.

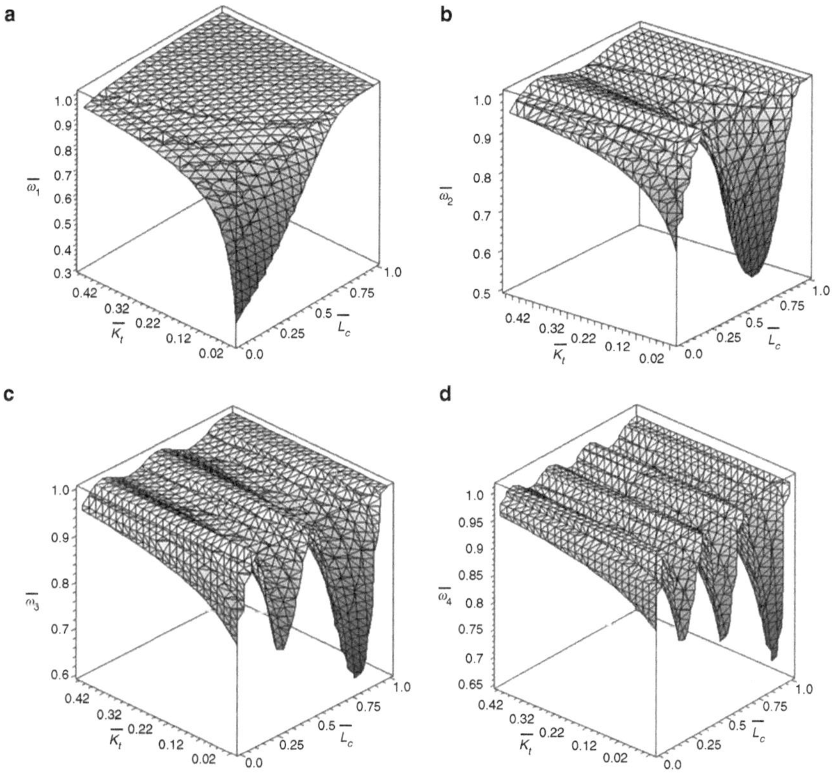

Fig. 6.20 Plots of change in the first four natural frequencies $\overline{\omega}_i$ of the transverse vibrations of rod on location $\bar{L}_c$ and rigidity $\bar{K}_t$ of elastic element

Points where there is no dependence on the natural frequencies at the existence of the elastic element are the points of the kink of the curve of the oscillation form. Since these points do not coincide for different forms of oscillations, it is necessary to have information about several frequencies for solving the inverse problem of reconstruction of defect parameters. Analysis of tabular data made it possible to reveal a number of features in the nature of the frequency dependencies $\overline{\omega}_i$:

(i) The largest drop in the natural oscillation frequencies occurs for three isolated ranges of the spread of the elastic element location: the first, rather narrow ($\bar{L}_c =$ 0.20–0.31) and then two wide ($\bar{L}_c = 0.45$–0.55, $\bar{L}_c = 0.73$–0.95); in the narrow range, the first natural frequency decreases a significant amount ($0.93\,\overline{\omega}_1$), and in the third and fourth decreases by $0.29\,\overline{\omega}_3$ and $0.22\,\overline{\omega}_4$, respectively. In the second wide interval, the fourth natural frequency decreases by 0.26 $\overline{\omega}_2$. In the third wide spread interval ($\bar{L}_c = 0.73$–0.95), the values of all four natural frequencies decrease by different values;

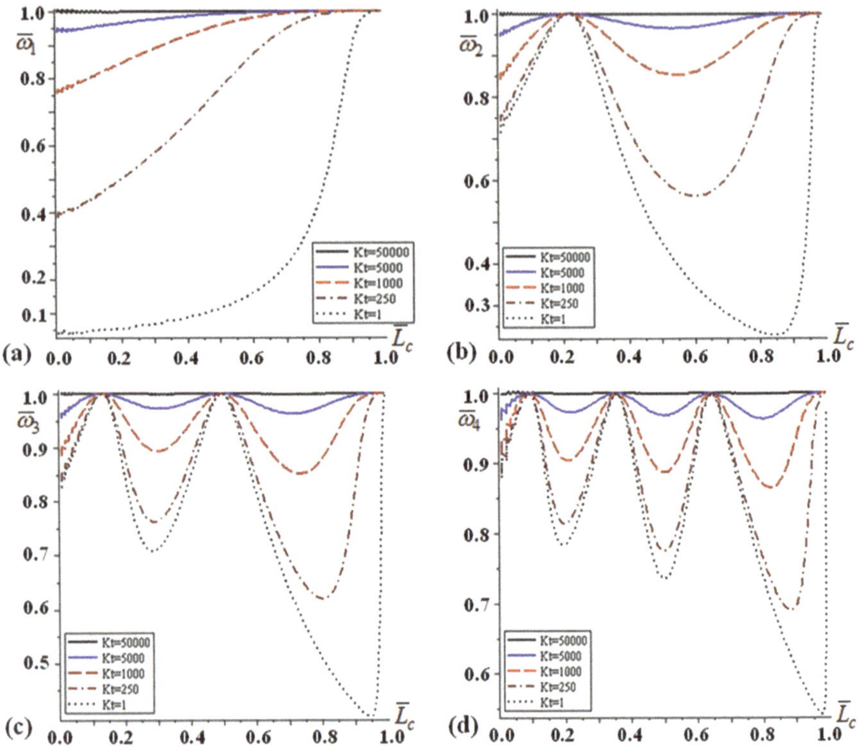

Fig. 6.21 Dependence of the relative change in the natural frequencies $\overline{\omega}_i$ of the flexural vibrations of rod on the location $\bar{L}_c$ and rigidity K_t (in N·m/rad) of elastic element: (**a**), (**b**), (**c**), and (**d**) correspond to first, second, third, and fourth natural frequencies

(ii) In addition to the above-mentioned peculiarities, near the pinching of the rod (at $\bar{L}_c = 0.02–0.05$), a small drop ($0.1–0.27\,\overline{\omega}_i$) was observed for the second, third, and fourth natural frequencies, while for the first mode, the frequency drops to $0.96\overline{\omega}_1$.

Another feature is the presence of "breaks" on the curves of the vibration modes. In regard to this, the vibration patterns are compared for the rigidities of the elastic element $K_t = 10{,}000$ N·m/rad and $K_t = 10$ N·m/rad at its location $\bar{L}_c = 0.25$. The natural frequencies $\overline{\omega}_i$ of the first four modes of oscillations are shown in Table. 6.12. For a given defect in the range $\bar{L}_c = 0.20–0.31$, a "break" in the shape of the oscillations is clearly observed on the first, third, and fourth modes and weakly on the 2nd mode of oscillation. There are no such "breaks" on the vibration modes of the rod without a defect (shown by the solid lines in Fig. 6.22).

Analysis of vibration modes (Fig. 6.22) showed that the antinode of the second natural frequency is located at the point $\bar{L}_{c(2)} = 0.48$, and for the third natural

Table 6.11 Location of elastic element at the greatest frequency variation for various natural frequencies

No. of natural frequency	Range of elastic element location $\bar{L}_c$ (value of relative decrease in natural frequency in the corresponding range)			
1	–	0–0.92/(>0.7 $\overline{\omega}_1$)	–	–
2	0–0.06/(0.2 $\overline{\omega}_2$)	–	0.57–0.77/(0.72 $\overline{\omega}_2$)	–
3	0–0.05 / (0.15 $\overline{\omega}_3$)	0.28–0.31/(0.29 $\overline{\omega}_3$)	–	0.73–0.92/(0.57 $\overline{\omega}_3$)
4	0–0.03/ (0.10 $\overline{\omega}_4$)	0.20–0.23/(0.22 $\overline{\omega}_4$)	0.45–0.55/(0.26 $\overline{\omega}_4$)	0.81–0.95/(0.45 $\overline{\omega}_4$)

Table. 6.12 Natural frequencies of the model for different modes of oscillations with two rigidities of elastic element at $\bar{L}_c = 0.25$

	Oscillation mode			
	1	2	3	4
Rigidity of elastic element K_t, N·m/rad	Natural frequencies of model ω_i, Hz			
10,000	107.2	676.2	1875	3672
10	21.3	139.4	1394	3114

frequency it is located at $\bar{L}_{c(3)}^{'} = 0.28$ and $\bar{L}_{c(3)}^{''} = 0.7$. For the fourth natural frequency, the antinodes are located at $\bar{L}_{c(4)}^{'} = 0.22$, $\bar{L}_{c(4)}^{''} = 0.5$, $\bar{L}_{c(4)}^{'''} = 0.78$.

Comparison of the range of $\bar{L}_c$, in which there is a significant drop in the natural frequencies of different oscillation modes, as well as the presence of "breaks" on the curves of the oscillation forms, allow us to conclude that this decrease in the natural frequencies is a sign of identifying the location of the defect in the rod. The fact that the maximum narrowing of three natural frequencies is observed in the first narrow range of $\bar{L}_c$ (Table 6.11) allows us to formulate an assumption on the method of identifying the location of one of the most dangerous defects in the rod (cantilever). It is based on the greatest degree of dropping the values of three or more natural frequencies of its oscillations.

When the sensitivity of the frequencies was determined from the elastic element (spring) stiffness, the cross-sections of the surfaces (Fig. 6.20) were considered for planes with different defect locations, which are a graphic representation of the solution of Eq. set (6.17) in the following formulation: $\bar{L}_c = \{0.05; 0.25; 0.4; 0.8\}$. The arrangement of the elastic element for calculations is shown in Fig. 6.23. It was chosen taking into account the sensitivity to a certain location $\bar{L}_c$ on the base of dependency $\overline{\omega}_i(\bar{L}_c)$.

Figure 6.24 shows the dependencies of the first four natural frequencies $\overline{\omega}_i$ ($i = 1$–4) on the rigidity K_t of the elastic element at its different locations $\bar{L}_c$. Analysis of the dependencies (Fig. 6.24) shows that the frequency variation from stiffness of elastic element at its different arrangements is different and monotonous. At high stiffness values $K_t > 2000$ N·m/rad, i.e. at small defect sizes $\bar{t}$, the frequency

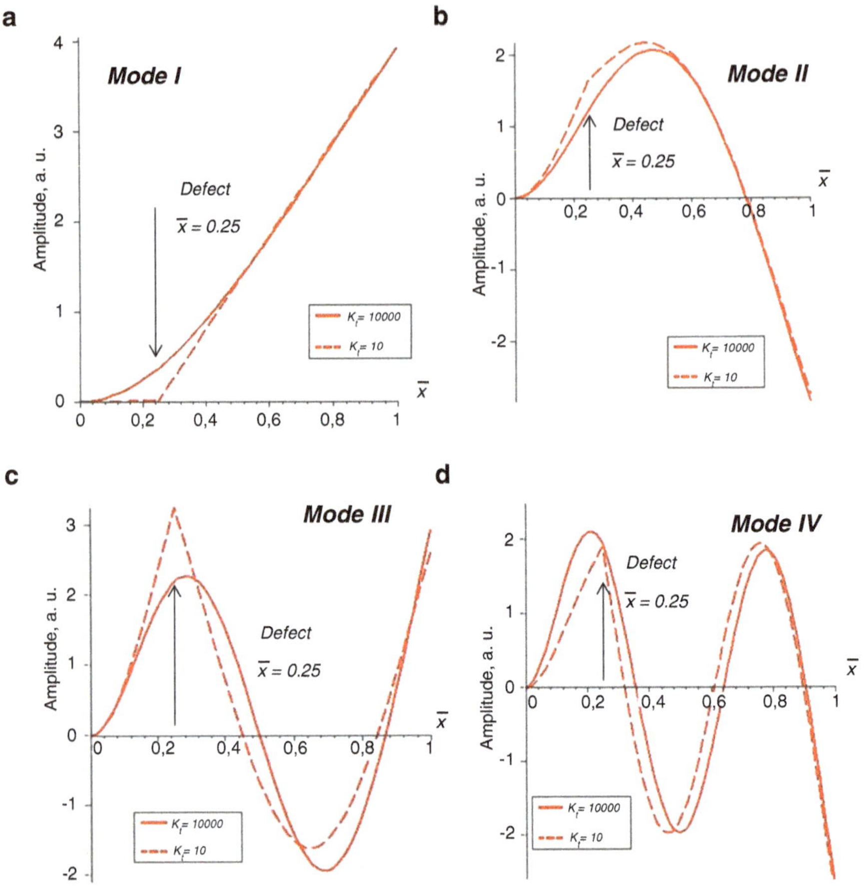

Fig. 6.22 Forms of different oscillation modes of the rod with elastic element with stiffness coefficients $K_t = 10{,}000$ N·m/rad (conditionally intact, represented by a solid line) and $K_t = 10$ N·m/rad (defective, dotted)

change is insensitive and can be within 10% for the first and second modes and 7% for the third and fourth modes of oscillations.

6.4 Methods of Identifying Defects in Cantilever

In Sect. 6.3.2 we considered the solution of the direct problem of finding the dependence of the natural frequencies on the defect stiffness K_t and its location $\bar{L}_C$. In practice, it is possible to obtain information about the state of a construction with a defect in the form of a set (spectrum) of natural frequencies. By using the analytical model described in Sect. 6.3 it is possible to solve the inverse problem of identifying

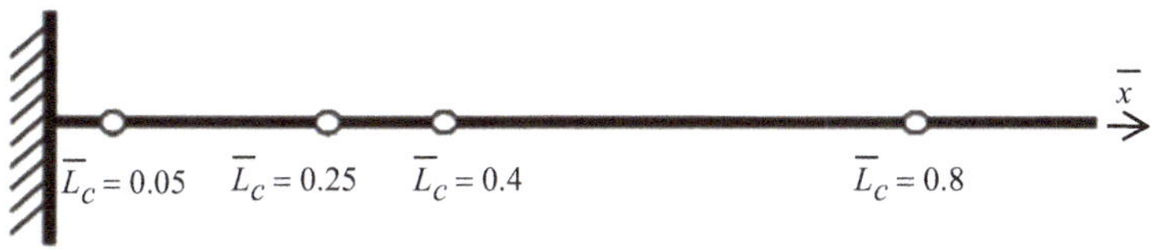

Fig. 6.23 Arrangement of the points of elastic element along rod

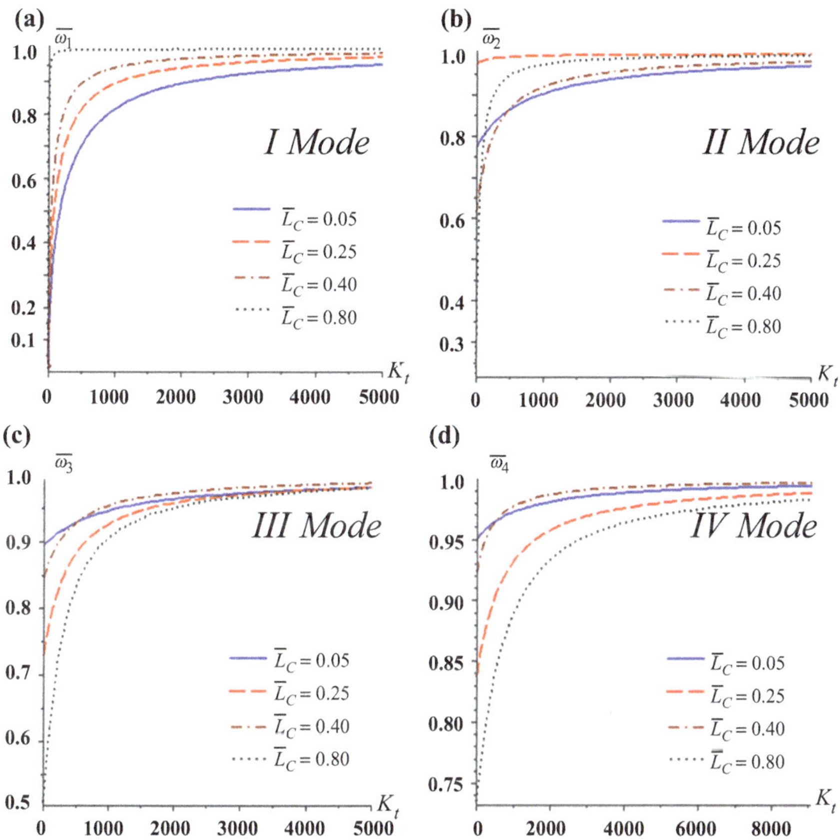

Fig. 6.24 Dependence of the rigidity K_t (in N·m/rad) of elastic element on the relative change of natural frequencies $\bar{\omega}_i$ for various locations $\bar{L}_c$

the defect parameters, that is the value of the rigidity K_t of the elastic element and its location $\bar{L}_C$.

In this section, we present the method of reconstruction of defects, based on the analytical solution of the rod model.

The first step consists of determination of the approximate relationship between the rigidity of the elastic element and the defect size $\bar{t}$ (or the moment of inertia I of the damaged section). To evaluate this dependence, we compare the analytical solution for the natural oscillations of the rod and the numerical solution of the

Table 6.13 The first four natural frequencies of transverse oscillations, depending on the location $\bar{L}_c$ and size $\bar{t}$ of the defect

	Defect location	Defect size	Natural frequencies (ω_i^*), based on FE modeling, Hz			
No. of variant	$\bar{L}_c$	$\bar{t}$	ω_1^*	ω_2^*	ω_3^*	ω_4^*
1	0.1	0.1	107,3	672	1,871	3,630
2	0.1	0.3	102	662	1,867	3,629
3	0.1	0.5	91.5	642	1,857	3,625
4	0.1	0.7	68	611	1,840	3,611
5	0.1	0.8	48	594	1,830	3,571
6	0.1	0.9	22.9	581	1,818	3,624
7	0.3	0.1	107	673	1,866	3,626
8	0.3	0.3	105.5	669	1,825	3,599
9	0.3	0.5	99.7	659	1,729	3,538
10	0.3	0.7	84	636	1,552	3,408
11	0.3	0.8	67	615	1,457	3,298
12	0.3	0.9	37	591	1,339	3,152
13	0.4	0.1	107	671	1,869	3,626
14	0.4	0.3	106	658	1,847	3,601
15	0.4	0.5	102	624	1,796	3,538
16	0.4	0.7	90	548	1,699	3,390
17	0.4	0.8	73	488	1,611	3,245
18	0.4	0.9	40.2	431	1,578	3,064
19	0.6	0.1	108	671	1,868	3,627
20	0.6	0.3	107.7	654.2	1,840	3,610
21	0.6	0.5	106.7	609	1,776	3,568
22	0.6	0.7	103.1	500.8	1,661	3,472
23	0.6	0.8	96.7	403.6	1,590	3,390
24	0.6	0.9	72.4	285.2	1,527	3,284
25	0.8	0.1	108.3	673	1,866	3,612
26	0.8	0.3	108.1	669	1,825	3,499
27	0.8	0.5	108	658	1,712	3,258
28	0.8	0.7	107.9	613	1,413	2,899
29	0.8	0.8	107	528	1,170	2,742
30	0.8	0.9	102	309	987	2,643

modal analysis problem for the full-body bar by using ANSYS software. Construction of the dependence expresses the dynamic equivalence of these models. In this case, an overdetermined system of Eq. (6.19) was solved with respect to one unknown (rigidity K_t of the elastic element), in the presence of information about location $\bar{L}_c$, size $\bar{t}$ of the defect, and natural frequencies $\omega_i^*(\bar{t})$, obtained by modeling the damaged rod by using ANSYS software:

$$\Delta\left(\omega_i^*(\bar{t}), K_t, \bar{L}_c\right) = 0; \; i = 1...4. \tag{6.19}$$

The second step includes the solution of an overdetermined system of Eq. (6.19) with respect to two unknowns (K_t and $\bar{L}_c$) by using the first four natural frequencies $\omega_i^*(\bar{t})$ of bending vibrations.

The third step is to find the size of the defect $\bar{t}$ or the moment of inertia I of the damaged section due to recalculation using the previously established dependencies $\bar{t}(K_t)$ or $I(K_t)$.

6.4.1 Comparison of Finite-Element and Analytical Models on the Base of Dynamic Equivalence

The problem of establishing an approximate relationship between the stiffness parameter K_t of elastic element in the analytical model and the defect size $\bar{t}$ as well as the moment of inertia I of the damaged section for the full-body FE model is solved.

With this aim, we consider the solution of the set of Eq. (6.19) in the presence of information on the location of the defect and the natural frequencies of the rod. To obtain the dependence $K_t = K_t(\bar{t})$ for a single defect, the first four natural frequencies are taken into account in the problem, which can be obtained quite accurately from the experiment. In the present section, this experiment was replaced by a calculation in the FE complex ANSYS. In the calculation, the oscillations in the plane of greatest rigidity are considered.

For different variants of the defect location and its size, the spectra of natural frequencies of the cantilever are calculated on the base of FE simulation by using ANSYS software (Table 6.13). Then the problem of determining the rigidity K_t^* of an elastic element was solved using Eq. (6.19) by using MAPLE software, in the presence of information on its location $\bar{L}_c$ and frequency spectrum ω_i^* . The results are shown in Table 6.14. Columns 2 and 3 of Table 6.14 give the corresponding variants of location $\bar{L}_c$ and size $\bar{t}$ of defect for the calculation of natural frequencies using the finite-element ANSYS software. The calculated values of the rigidity K_t^* of the elastic element are present in column 4.

Stiffness values K_t^* in the dependence on the defect size $\bar{t}$ for different locations $\bar{L}_c$ of the elastic element for a full-body model are shown in Table 6.15.

To determine the relationship between the defect size $\bar{t}$ in the FE full-body model and the rigidity K_t^* of the elastic element in the analytical model, a correlation analysis was performed, taking into account the various locations $\bar{L}_c$ of the defect. Analysis of the data for different stiffness K_t^* values at the same defect sizes shows that the deviation from the average calculated value of K_t^* is from 5% to 21%. The greatest deviation is achieved with small stiffness K_t^* values.

Table 6.14 Calculation results of defect rigidity $K_t{}^*$ based on analytical modeling

	Given parameters		Solution of inverse problem (Maple software)
No. of variant	Defect location, $\bar{L}_c$	Defect size, $\bar{t}$	Calculated rigidity, K_t^*, N·m/rad
1	0.1	0.1	9,500
2	0.1	0.3	3,636
3	0.1	0.5	1,106
4	0.1	0.7	283
5	0.1	0.8	106
6	0.1	0.9	20
7	0.3	0.1	9,252
8	0.3	0.3	3,753
9	0.3	0.5	1,112
10	0.3	0.7	303
11	0.3	0.8	124
12	0.3	0.9	26.5
13	0.4	0.1	8,157
14	0.4	0.3	3,284
15	0.4	0.5	1,019
16	0.4	0.7	277
17	0.4	0.8	104
18	0.4	0.9	20
19	0.6	0.3	8,135
20	0.6	0.3	2,742
21	0.6	0.5	1,054
22	0.6	0.7	305
23	0.6	0.8	124
24	0.6	0.9	27.7
25	0.8	0.1	8,135
26	0.8	0.3	3,514
27	0.8	0.5	1,039
28	0.8	0.7	263
29	0.8	0.8	96
30	0.8	0.9	17.7

The approximate dependence $\bar{t}(K_t)$ was sought in the form:

$$\bar{t} = a + b(K_t{}^*)^n, \tag{6.20}$$

where a, b, and n are the sought values of the dependence.

By using the method of least squares, the following expression was attained:

Table 6.15 Dependence of defect size $\bar{t}$ on different values of stiffness K_t^* and location $\bar{L}_c$ of the elastic element

	Defect size $\bar{t}$					
	0.1	0.3	0.5	0.7	0.8	0.9
Location of elastic element, $\bar{L}_c$	Stiffness of elastic element, K_t^*, N·m/rad					
0.1	9,500	3,636	1,106	283	106	20
0.3	9,252	3,753	1,112	303	124	25.5
0.4	8,157	3,284	1,019	277	104	20
0.6	8,135	2,742	1,054	305	124	26
0.8	8,121	3,514	1,039	263	96	17.7

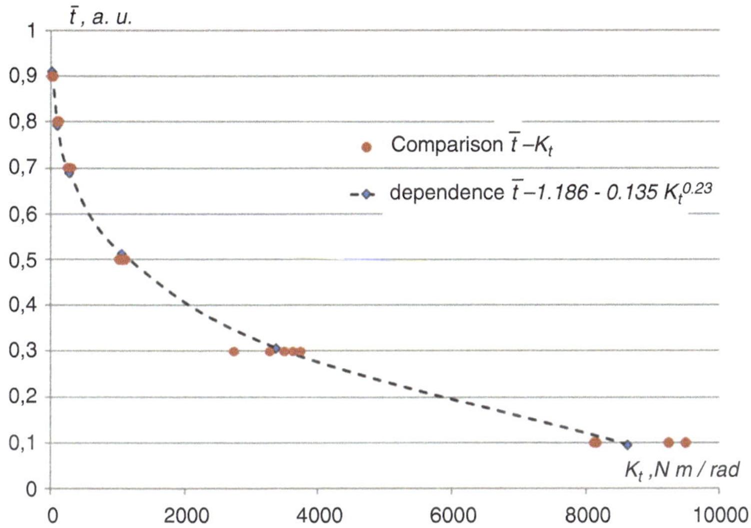

Fig. 6.25 Dependence between the defect size $\bar{t}$ and the rigidity K_t of elastic element for different its locations $\bar{L}_c$ along the rod length

$$\bar{t} = 1.186 - 0.135\left(K_t^*\right)^{0.23} \tag{6.21}$$

with correlation factor $R = 0.97$.

Graphic interpretation of this dependence is also present in Fig. 6.25. The plot of the dependence has a monotone decreasing character.

The results of the calculated defect sizes $\bar{t}^*$ for some variants of locations $\bar{L}_c$ of the defect are present in Table 6.16 in column 4. The deviation $\Delta\bar{t}$ in determining the defect value with respect to the given one is calculated by formula (6.22) and presented in column 5:

$$\frac{\Delta\bar{t} = \left(\bar{t}^* - \bar{t}\right)}{\bar{t}100\%} \tag{6.22}$$

Table 6.16 Calculated defect size $\bar{t}$ for different locations $\bar{L}_c$

	Given defect location	Given defect size	Calculated defect size	Deviation from given value, %
No. of variant	$\bar{L}_c$	$\bar{t}$	$\bar{t}^*$	$\Delta\bar{t}$
3	0.4	0.1	0.11	10.00
5	0.8	0.1	0.11	10.00
8	0.4	0.3	0.31	3.33
10	0.8	0.3	0.31	3.33
12	0.3	0.5	0.51	2.00
16	0.1	0.7	0.69	−1.43
18	0.4	0.7	0.69	−1.43
21	0.1	0.8	0.79	−1.25
24	0.6	0.8	0.78	−2.50
26	0.1	0.9	0.92	2.22
27	0.3	0.9	0.90	0.00
30	0.8	0.9	0.92	2.22

Moreover, a relationship was found between the moment of inertia $I(K_t)$ of the cross-section of full-body rod and the rigidity K_t of the elastic element of analytical model. In the calculation, the moment of inertia of the cross-section relative to the principal axis passing through the center of the damaged element in the cross-section was considered.

The moment of inertia I of a section with a defect in the form of a notch for a rectangular cross-section, depending on the rigidity K_t of the elastic element of the analytical model, can be described by the following equation:

$$I(K_t) = \frac{bh^3[1 - \bar{t}(K_t)]^3}{12}, \tag{6.23}$$

where h and b are the absolute values of the rod height and width.

By substituting Eq. (6.21) into Eq. (6.23), the approximate dependence of the moment of inertia on the rigidity of the elastic element can be described as

$$I(K_t) = \frac{bh^3\left(0.135K_t{}^{0.23} - 0.186\right)^3}{12}. \tag{6.24}$$

A graphic interpretation of the curve of the approximate dependence is shown in Fig. 6.26. The correlation coefficient is $R = 0.96$.

6.4.2 Reconstruction of Defect Parameters in Cantilever

By reconstructing the defect parameters on the base of using an analytical model, an overdetermined set of Eq. (6.19) was solved with respect to two unknowns: K_t and

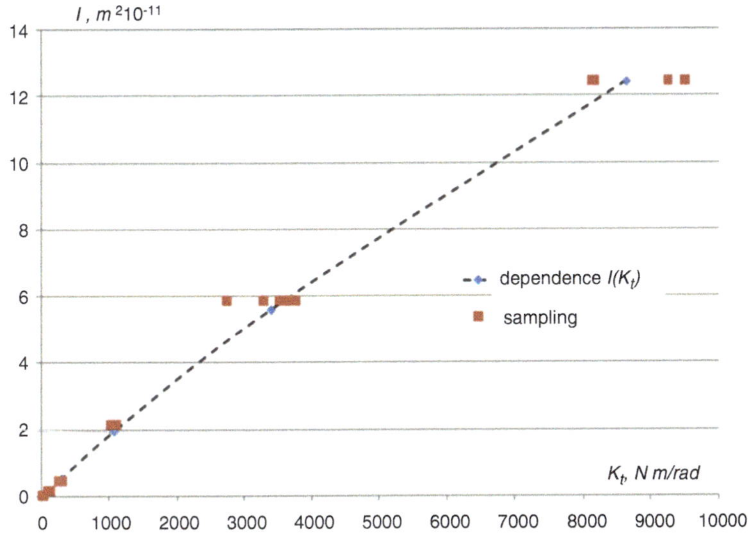

Fig. 6.26 Dependence of the inertia moment I of a section with a defect for full-body model on the rigidity K_t of elastic element in analytical model

$\bar{L}_c$. As the input data, the first four natural frequencies ω_i^* of the flexural oscillations, obtained during the modeling by using ANSYS software, were used. These frequencies can also be obtained in the result of the full-scale experiment, which will be described in Chap. 7.

One of the ways to find unknown parameters K_t and $\bar{L}_c$ from Eq. (6.19) consists of solving the problem of minimizing the discrepancy of the following set of equations:

$$\sum\nolimits_{i=1}^{k} \left|\Delta\left(\omega_i^*, K_t, \bar{L}_c\right)\right| \to \min, \quad i = 1, ..k \tag{6.25}$$

As an example of determining the location $\bar{L}_c$ and rigidity K_t of an elastic element (defect) in calculation, a model with a defect size of $\bar{t} = 0.5$ and its location $\bar{L}_c = 0.4$ was considered. The first four natural frequencies were calculated on the base of the modal analysis in the finite-element ANSYS software:$\omega_1^* = 102$ Hz, $\omega_2^* = 624$ Hz, $\omega_3^* = 1{,}796$ Hz, $\omega_4^* = 3{,}599$ Hz. Based on the frequencies obtained, a system of Eq. (6.19) was solved to determine the notch stiffness K_t and its location $\bar{L}_c$. A graphic interpretation of the constructed dependencies $\Delta\left(\omega_i^*, K_t, \bar{L}_c\right) = 0$ is shown in Fig. 6.27 for the previously presented model.

The point in the circle in Fig. 6.27 corresponds to a general solution that satisfies all four equations of frequency determinants. It should be noted that with a certain degree of certainty it can be assumed that the intersection of curves 1, 2, 3, and 4 of the dependence $K_t\left(\bar{L}_c\right)$ (Fig. 6.27) with the coordinates $\{K_t = 957$ N·m/rad, $\bar{L}_c = 0.4\}$ corresponds to the location of the defect (notch). By recalculating the rigidity of an elastic element by using formula in Eq. (6.21), the defect size $\bar{t} = 0.53$ is found, and the deviation from the given value equals 6%.

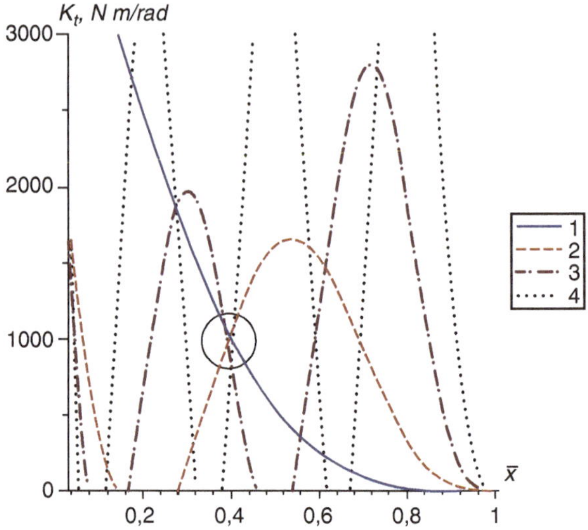

Fig. 6.27 Frequency determinants of system: 1 – $\Delta(\omega_1{}^*, K_t, \bar{L}_c) = 0$; 2 – $\Delta(\omega_2{}^*, K_t, \bar{L}_c) = 0$; 3 – $\Delta(\omega_3{}^*, K_t, \bar{L}_c) = 0$; 4 – $\Delta(\omega_4{}^*, K_t, \bar{L}_c) = 0$

In Sect. 6.4.1, the stiffness parameter was determined in the presence of input data on the natural frequencies given in Table 6.13. At this stage, the location of the defect and the error in its determination are considered by comparing the analytical and finite element models. The calculation results are given in Table 6.17. Columns 2 and 3 of Table 6.17 give the appropriate variants of defect location $\bar{L}_c$ and size $\bar{t}$ for the calculation of natural frequencies in the finite-element ANSYS software. The values of location $\bar{L}_c{}^*$ and stiffness $K_t{}^*$ found are given in columns 4 and 6, respectively. The deviation $\Delta\bar{L}_c$ of the calculated values $\bar{L}_c{}^*$ from the given values $\bar{L}_c$ is determined by the following formula:

$$\Delta L_C = \frac{\left(\bar{L}_C^* - \bar{L}_C\right)}{\bar{L}_C} \times 100\%. \tag{6.26}$$

By determining the defect location $\bar{L}_c{}^*$, the deviation reaches 12.9%. The maximum error is reached for cases with high rigidity, and, thereby, the minimum size $\bar{t}$ of the defect.

6.5 Investigation of the Features of Resonance Modes of Cantilever with Defect

6.5.1 Comparison of Oscillation Modes of FE and Analytical Models

We perform comparison in this section of the oscillation forms obtained by numerical modeling in the FE ANSYS software (Sect. 6.2) and the analytical calculation of the rod simplified model with an elastic element (Sect. 6.3.2).

Table 6.17 Results of calculating the defect location $\bar{L}_c{}^*$ and stiffness $K_t{}^*$, based on analytical modeling

No. of variant	Given parameters		Solution of inverse problem (Maple software)		
	Defect location	Defect size	Calculated defect location	Error, %	Calculated rigidity
	$\bar{L}_c$	$\bar{t}$	$\bar{L}_c{}^*$	$\Delta\bar{L}_c$	K_t^*, N·m/rad
1	0.1	0.1	0.094	−6.0	9,500
2	0.1	0.3	0.0911	−8.9	3,636
3	0.1	0.5	0.095	−5.0	1,106
4	0.1	0.7	0.097	−3.0	283
5	0.1	0.8	0.098	−2.0	106
6	0.1	0.9	0.097	−3.0	20
7	0.3	0.1	0.334	11.3	9,252
8	0.3	0.3	0.324	8.0	3,753
9	0.3	0.5	0.309	3.0	1,112
10	0.3	0.7	0.303	1.0	303
11	0.3	0.8	0.303	1.0	124
12	0.3	0.9	0.303	1.0	26.5
13	0.4	0.1	0.36	−10.0	8,157
14	0.4	0.3	0.397	−0.8	3,284
15	0.4	0.5	0.399	−0.3	1,019
16	0.4	0.7	0.402	0.5	277
17	0.4	0.8	0.4	0.0	104
18	0.4	0.9	0.4	0.0	20
19	0.6	0.3	0.671	12.9	8,135
20	0.6	0.3	0.674	12.3	2,742
21	0.6	0.5	0.623	3.8	1,054
22	0.6	0.7	0.605	0.8	305
23	0.6	0.8	0.603	0.5	124
24	0.6	0.9	0.6	0.0	27.7
25	0.8	0.1	0.788	−1.5	8,135
26	0.8	0.3	0.768	−4.0	3,514
27	0.8	0.5	0.79	−1.3	1,039
28	0.8	0.7	0.802	0.3	263
29	0.8	0.8	0.833	4.1	64.9
30	0.8	0.9	0.81	1.3	17.7

Modal calculation is conducted for models with defect location $\bar{L}_c = 0.25$. The models with two defect sizes ($\bar{t} = 0.3$ and $\bar{t} = 0.7$) were analyzed. In recalculation, applying the formula from Eq. (6.21), the flexural rigidity $K_t(\bar{t})$ of the elastic element for the analytical model was: $K_t(0.7) = 262$ N·m/rad and $K_t(0.3) = 3569$ N·m/rad.

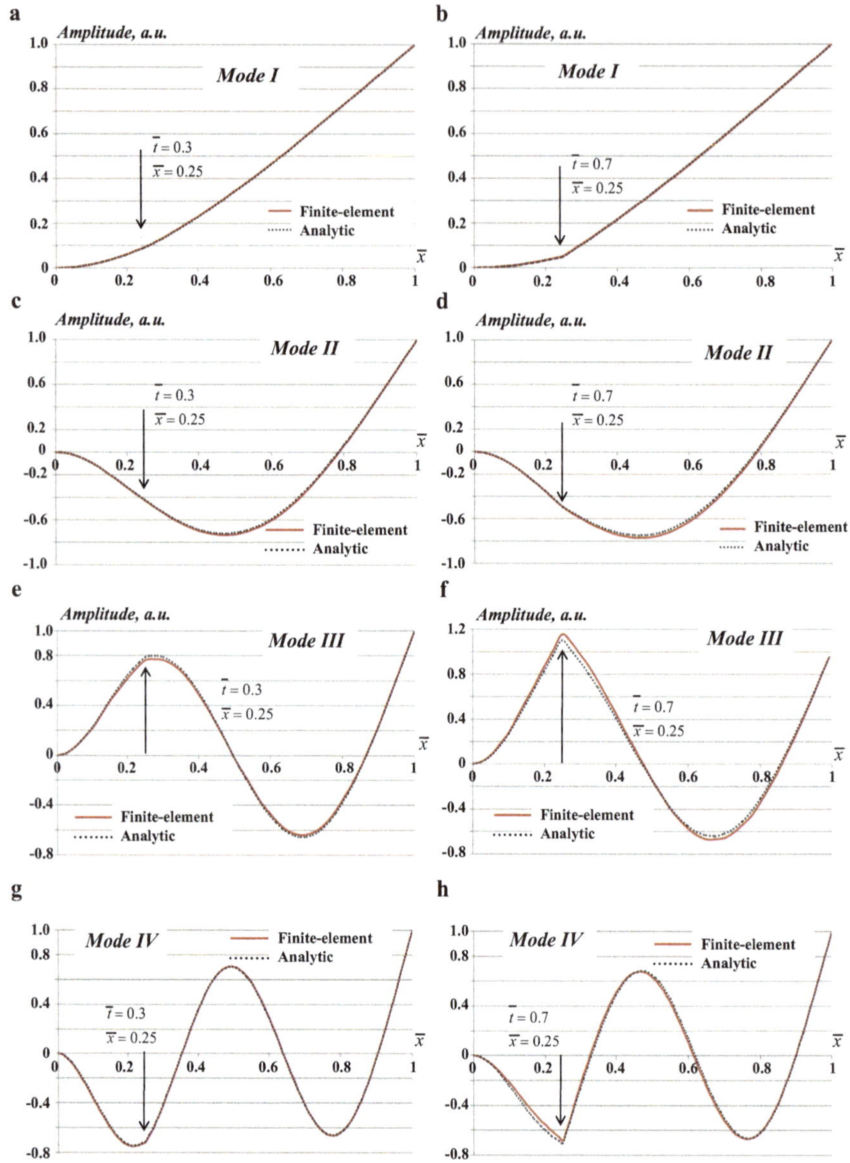

Fig. 6.28 Forms of the first (**a**, **b**), second (**c**, **d**), third (**e**, **f**), and fourth (**g**, **h**) modes of transverse oscillations, obtained in the result of FE and analytical modal calculations for defect size $\bar{t} = 0.3$ (**a**, **c**, **e**, **g**) and $\bar{t} = 0.7$ (**b**, **d**, **f**, **h**)

Figure 6.28 presents forms of the first (a, b), second (c, d), third (e, f), and fourth (g, h) modes of transverse oscillations, obtained with the result of FE and analytical calculations for the quantities of defect $\bar{t} = 0.3$ (a, c, e, g) and $\bar{t} = 0.7$ (b, d, f, h). To compare the vibration forms, the amplitudes were normalized to the amplitude of the oscillations at the point on the free edge of the rod $\bar{x} = 1$.

At the location of the defect, there is a break in the shape of oscillations, which is clearly manifested at the third and fourth modes of oscillations for the defect size $\bar{t} = 0.3$ and on all selected forms of the oscillation modes at the defect size $\bar{t} = 0.7$ for both models. The oscillation forms, obtained on the base of the analytical calculation, were compared with the vibration forms, obtained by the FE method, for each point along the length of the rod. The relative divergence of the amplitudes of the oscillation forms at various points along the length of the rod was calculated as follows:

$$\Delta\bar{A} = \frac{\left|\bar{A}_{\text{an}} - \bar{A}_{\text{FE}}\right|}{\bar{A}_{\max\ \text{FE}}} \times 100\%. \tag{6.27}$$

The comparative analysis of the corresponding curves of the vibration modes showed the following. When the amplitudes of the first oscillation mode are compared at the corresponding points along the length of the rod, the maximum discrepancy $\Delta\bar{A}_{\max} = 0.2\%$ takes place at the defect size $\bar{t} = 0.3$ and $\Delta\bar{A}_{\max} = 1.2\%$ at the defect size $\bar{t} = 0.7$. In this case, the maximum amplitude divergence lies near the defect location.

When we compare the amplitudes of the first oscillation mode at the corresponding points along the rod length, the maximum discrepancy $\Delta\bar{A}_{\max} = 1.47\%$ for defect size $\bar{t} = 0.3$ and $\Delta\bar{A}_{\max} = 2.61\%$ for defect size $\bar{t} = 0.7$. The maximum discrepancy between the amplitudes of the vibration modes is observed at the point of breaking the vibration modes ($\bar{x} = 0.61$). Near the defect location, for two variants of the defect, the difference in the amplitudes of the second oscillation form does not exceed $\Delta\bar{A}_{\max} = 0.39\%$.

A comparison of the amplitudes of the third oscillation mode at the corresponding points along the rod length shows that the maximum amplitude divergence $\Delta\bar{A}_{\max} = 2.7\%$ and occurs at a defect size $\bar{t} = 0.3$ and $\Delta\bar{A}_{\max} = 6.5\%$ at $\bar{t} = 0.7$. The maximum discrepancies of the amplitudes correspond to the points of the rod near the defect location for its two variants of sizes.

A comparison of the amplitudes of the fourth oscillation mode at the corresponding points along the rod length shows that the maximum amplitude discrepancy $\Delta\bar{A}_{\max} = 1.49\%$ takes place at a defect size $\bar{t} = 0.3$ and $\Delta\bar{A}_{\max} = 4.7\%$ at $\bar{t} = 0.7$. The maximum discrepancies correspond to the amplitudes near the defect location points.

Analysis of the compared forms of vibrations for different defect sizes shows that the qualitative characteristics of the curves of the vibration modes both near the location of the defect and along the length of the rod are the same for both models.

6.5.2 Choice of Characteristics for Identification of Defects in Cantilever, Based on the Analysis of Eigen-Forms of Bending Oscillations

This section substantiates the diagnostic features that characterize the defect location and its size for a cantilever-pinched elastic rod, based on the analysis of the features of oscillation forms. Here, indirect signs of defect identification are considered with the help of analysis of the change in the angle of tangents or the curvature of the dependence of bending angles and the curvature of the dependence of the oscillation modes of rod on the parameters of the defect.

By comparing the vibration modes with different stiffness values of the elastic element, as shown in Sect. 6.3.2, there is a sharp change in the angle between the tangents, demonstrated by the "break" in the shape of the oscillations at the point of the defect location. As an indicative characteristic of the presence of a defect in the cantilever rod, an angle α at a point, formed by tangents to the oscillation form curve could be used.

The angle α between the tangents at the points of the oscillation shape can be calculated using the discrete approach associated with the process of measuring the amplitudes of oscillations at a finite number of points. Figure 6.29 shows the scheme of the location of points on the section of the oscillation form curve:

In the presence of discrete information on the shape of the oscillations (Fig. 6.29), the magnitude of the angle at a point i for the corresponding mode of oscillation can be calculated as follows:

$$\alpha_i = \arccos\left(\frac{(\overline{AB})(\overline{BC})}{\|\overline{AB}\|\|\overline{BC}\|}\right), \tag{6.28}$$

where $\overline{AB}$ and $\overline{BC}$ is the vector representation of two segments, respectively, between the points of the normalized form of oscillations with numbers $[i-1, i]$ and $[i, i+1]$.

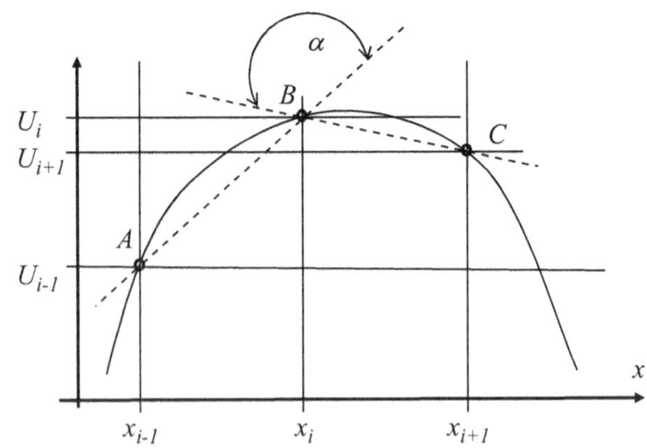

Fig. 6.29 Scheme of the location of points at calculation of α between tangents; U_i is the displacement of the natural oscillation mode at the i-th point of the rod; x_i is the coordinate of the point with the number i, $i \in 1, .., N$; N is the total number of points

The curvature of the oscillation shape can be considered as an additional feature that makes it possible to clarify the parameters of the defect.

When discrete measurements are used taking into account small oscillations, the curvature at the i-th point of the rod can be calculated by the following formula:

$$U_i'' = \frac{U_{i-1} - 2U_i + U_{i+1}}{\Delta x^2}, \tag{6.29}$$

where Δx is the distance between the measurement points.

When data are received and processed, it is necessary to collect the amplitudes of the oscillations to organize the vectors of the oscillation forms at different frequencies. For this, it is necessary to normalize the data on the amplitudes to the range [0, 1]. Each value of amplitude at the point of the oscillation form is normalized to the maximum deflection value:

$$\bar{U}_i = \frac{U_i}{|U_{\max}|}, \tag{6.30}$$

where $\bar{U}_i$ is the normalized value of the displacement at the i-th point of the rod; and $U_{\max}$ is the maximum deviation of the points of oscillation form.

Curvature at a point i at normalized amplitude of the shape of the oscillations will be calculated as:

$$\bar{U}_i'' = \frac{\bar{U}_{i-1} - 2\bar{U}_i + \bar{U}_{i+1}}{\Delta x^2}. \tag{6.31}$$

In Chap. 7, we consider a procedure of finding the angles between the tangents and the curvatures of the oscillation form based on experimental data.

6.5.3 Identification of Cantilever Defect Parameters, Based on the Analysis of Eigen-Forms of Bending Oscillations

In this section, we look at modeling the process of identifying the parameters of the cantilever defect, based on the analysis of the shapes of the first four modes of oscillation. The solution to the problem of finding the resonance frequencies and constructing the eigen-forms of the cantilever oscillations based on the analytical approach are considered. This approach can be applied as a substitution of the full-scale experiment.

In the analysis, we considered the variant with the location of the elastic element $\bar{L}_c = 0.25$. The stiffness K_t of the elastic element assumed values corresponding to the defect sizes of the full-body model with the following values: $\bar{t} = 0$; 0.25; 0.5; 0.75; 0.85. By recalculating and applying Eq. (6.21), the flexural rigidity $K_t(\bar{t})$ of the elastic element for the analytical model was equal to: $K_t(0) = 12684$ N·m/rad;

Table 6.18 Natural frequencies ω_i of cantilever for different values of the stiffness K_t of elastic element and its location $\bar{L}_c = 0.25$

	Natural frequencies, Hz			
K_t N·m/rad	ω_1	ω_2	ω_3	ω_4
12,684	107.8	676	1,892	3,709
4531	105.1	675.9	1,856	3,633
1173	98.1	674	1,772	3,486
163	67.9	667	1,533	3,216
52	44.9	664	1,440	3,147

$K_t(0.25) = 4531$ N·m/rad; $K_t(0.50) = 1173$ N·m/rad; $K_t(0.75) = 163$ N·m/rad; $K_t(0.85) = 52$ N·m/rad.

To find the proper modes of vibration for different values of the rigidity of the elastic element and its location $\bar{L}_c = 0.25$, based on the analytical modeling, the natural frequencies were calculated in the Maple software (Table 6.18).

The problem of the natural oscillations of cantilever was solved and forms of oscillations of the cantilever are obtained for various values of stiffness of the defect. The curvature values at various points are calculated from Eq. (6.31) and the angles between the tangents at different points along the cantilever length from Eq. (6.28). The length of the discrete segment Δx was taken as being equal to the 1/60 of the cantilever length.

Figure 6.30 presents normalized eigen-forms of the rod oscillations for different defect sizes. The plots of curvature at various points of the oscillation shape are also represented (Fig. 6.31) as well as the angles between the tangents at different points of the oscillation form curve (Fig. 6.32).

The analysis of the parameters of oscillation forms shows that at the defect location there is proper "breaking" defined by the defect size. Moreover, the analysis of the plots of curve and angles between the tangents at the defect location demonstrates clearly observed "peak." In order to estimate an influence of notch depth on changes of amplitude, oscillation form curvature, and angle in the zone of "breaking," we considered the dependencies of relative values of the parameters on stiffness of elastic element at the point of its location.

For a given case of the elastic element location, the relative value of normalized amplitude $\Delta\bar{U}$ is defined as

$$\Delta\bar{U}_i = \frac{\left|\bar{U}_i^d - \bar{U}_i^0\right|}{\bar{U}_i^0} \times 100\%, \tag{6.32}$$

where $\bar{U}_i^d$ and $\bar{U}_i^0$ are the normalized amplitudes of the oscillation form curves of the elastic rod at i-th along the rod length at the presence of the defect and in a non-damaged state, respectively. The values of transverse displacements for various oscillation Modes and their relative values at the point of defect location are present in Table 6.19.

Graphic interpretation of the dependence of the relative amplitude on the defect size $\bar{t}$ is shown in Fig. 6.33.

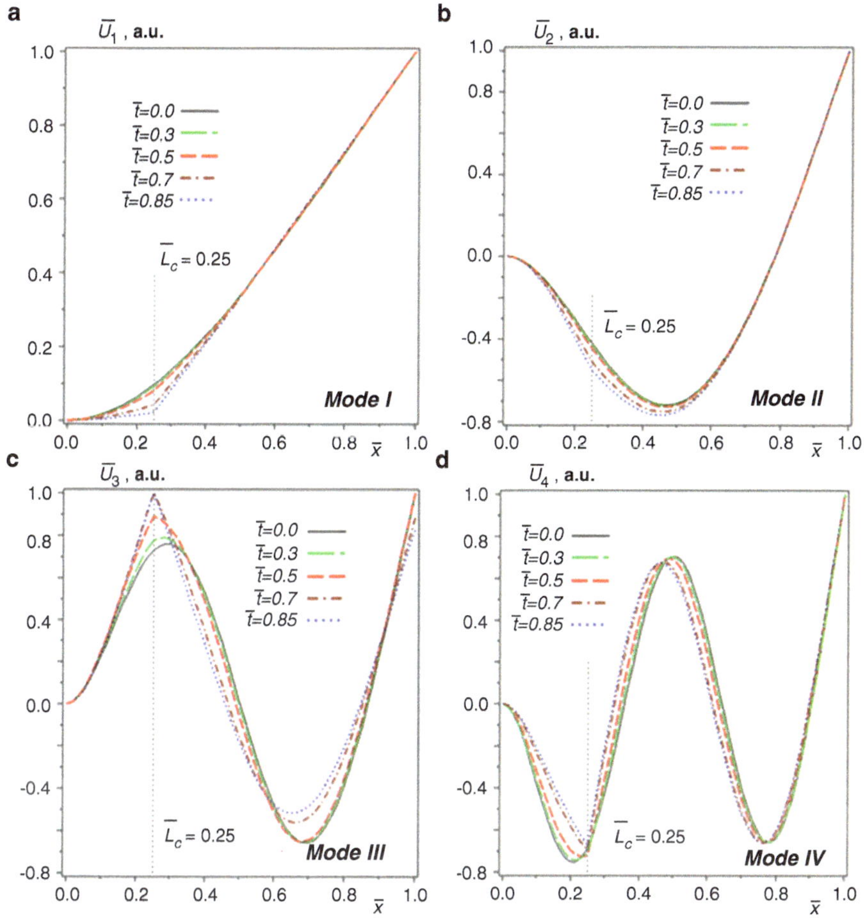

Fig. 6.30 Normalized values of the transverse displacements $\bar{U}$ of cantilever with different defects for the first four transverse oscillation modes. (**a**) first; (**b**) second; (**c**) third; (**d**) fourth

To perform a comparison, the curvature of the oscillation shape for this case of the arrangement of the elastic element was found as

$$\bar{U}''_i = \left|\bar{U}''_{id}\right|, \tag{6.33}$$

where $\bar{U}''_{id}$ is the value of the curvature of the oscillation shape at the i-th point along the rod length.

The values of the curvature of the oscillation form at the point where the defect is located are presented in Table 6.20.

Graphic interpretation of the dependence of the curvature U'' on the defect size $\bar{t}$ is shown in Fig. 6.34.

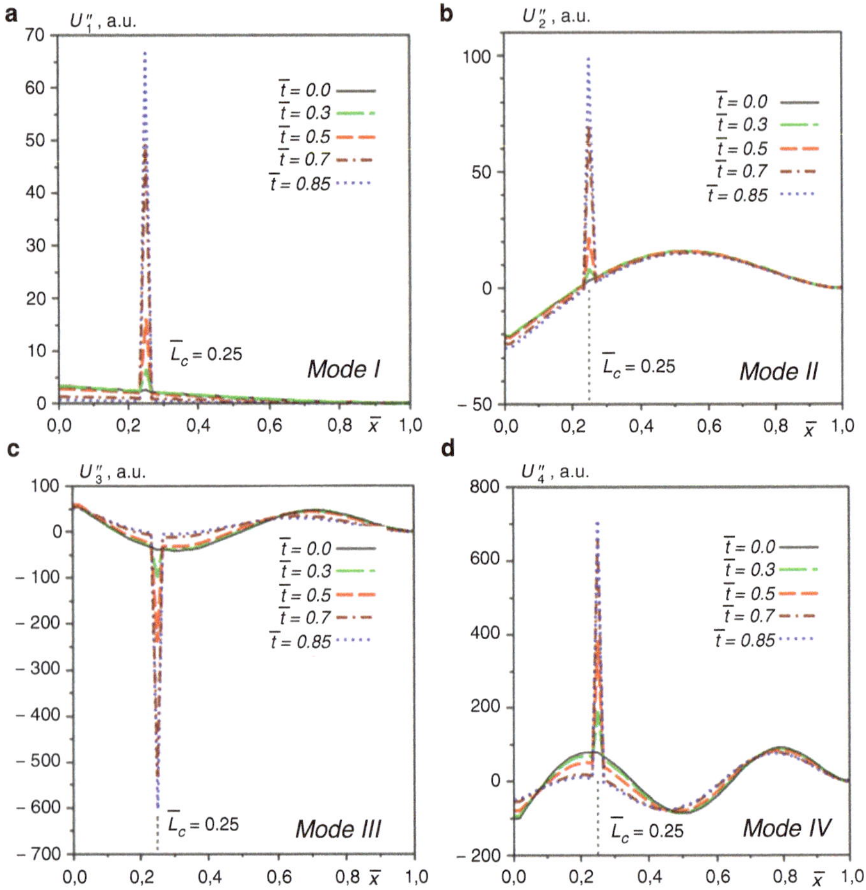

Fig. 6.31 Curvature U'' of the plot for the first four forms of transverse oscillations of cantilever with different defects: (**a**) first; (**b**) second; (**c**) third; (**d**) fourth

The relative angle between the tangents at the point of the oscillation form for the given case of the arrangement of the elastic element was calculated as follows:

$$\Delta\alpha_i = \frac{\alpha_i^d - \alpha_i^0}{\alpha_i^0} \times 100\%, \tag{6.34}$$

where α_i^d and α_i^0 are the angles at i-th point of the rod oscillation form at presence and absence of defect, respectively.

The angles of the "break" of the oscillation form and their relative values at the point of the defect location are shown in Table 6.21.

Graphic interpretation of the dependence of the relative value of angle between the tangents of oscillation forms at the point of defect location on the defect size $\bar{t}$ is shown in Fig. 6.35.

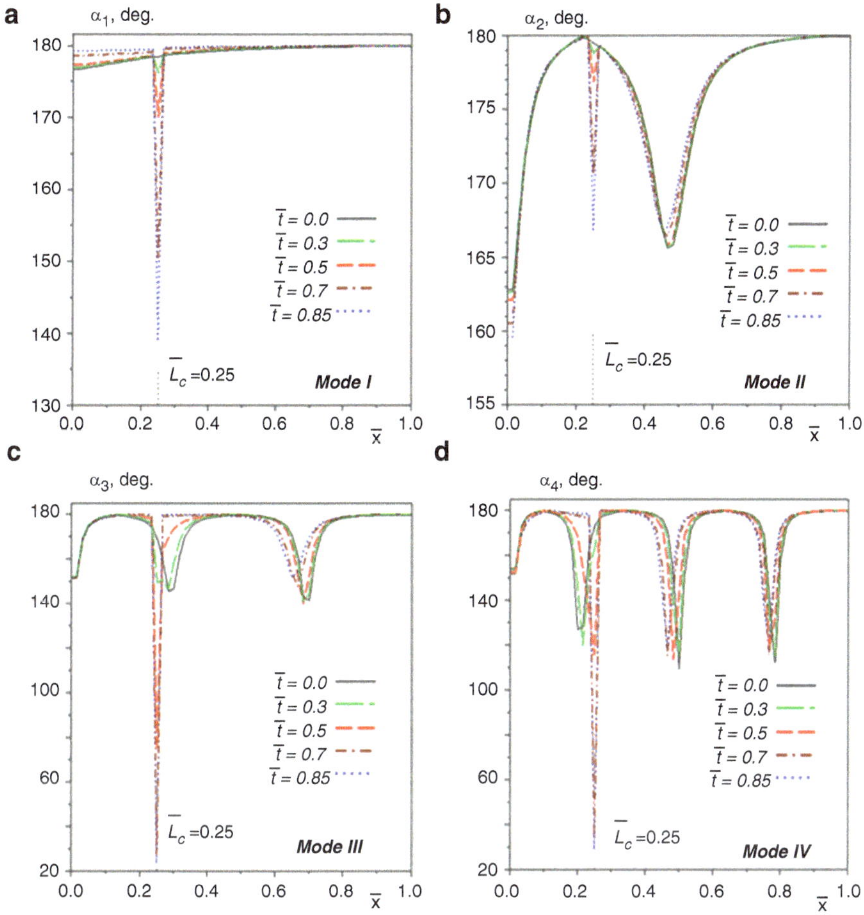

Fig. 6.32 Angles between the tangents at different points of the oscillation form curve for a cantilever with defects of various size $\bar{t}$ and for the first four transverse oscillation modes: (**a**) first; (**b**) second; (**c**) third; (**d**) fourth

Table 6.19 Transverse displacements $\bar{U}$ and their relative values $\Delta\bar{U}$ at the point of defect location for different oscillation modes of independence on stiffness K_t of cantilever

K_t, N·m/rad	$\bar{t}$, a. u.	Mode I $\bar{U}$, a. u..	Mode I $\Delta\bar{U}$, %	Mode II $\bar{U}$, a. u..	Mode II $\Delta\bar{U}$, %	Mode III $\bar{U}$, a. u..	Mode III $\Delta\bar{U}$, %	Mode IV $\bar{U}$, a. u.	Mode IV $\Delta\bar{U}$, %
12,684	0	0.097	0.0	−0.417	0.0	0.727	0.0	−0.686	0.00
4,531	0.25	0.093	4.1	−0.424	1.7	0.774	6.5	−0.703	2.48
1,173	0.50	0.081	16.5	−0.444	6.5	0.889	22.3	−0.723	5.39
163	0.75	0.0431	55.6	−0.51	22.3	1	37.6	−0.695	1.31
52	0.85	0.022	77.3	−0.551	32.1	1	37.6	−0.67	2.33

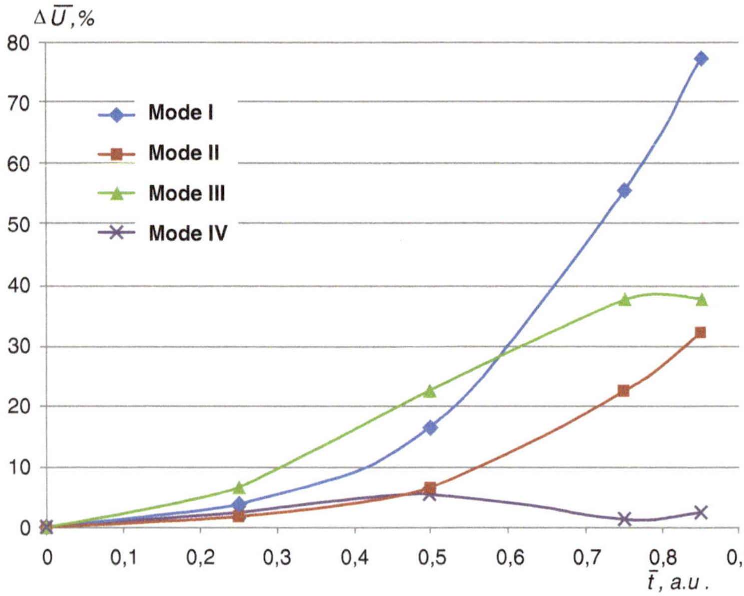

Fig. 6.33 Dependence of relative transverse displacement $\Delta\bar{U}$ at defect location site on defect size $\bar{t}$ for different oscillation modes

Table 6.20 Curvature of oscillation shape U'' at the point of defect location for different oscillation modes, depending on the cantilever rigidity K_t ($\bar{t}$ is the defect size)

		Mode I	Mode II	Mode III	Mode IV
K_t, N·m/rad	$\bar{t}$, a. u.	$\bar{U}''$, a. u.	$\bar{U}''$, a. u.	$\bar{U}''$, a. u.	$\bar{U}''$, a. u.
12,684	0	2.5	3.2	38.8	80.8
4,531	0.25	6.4	8.2	97.7	187
1,173	0.50	15.9	21.1	237	381
163	0.75	49.3	70.3	529	660
52	0.85	66.6	98.8	602	713

Analysis of the plots of oscillation forms, angles at points between tangents, and curvature shows that for the location $\bar{L}_c = 0.25$ of the elastic element, the relative change in displacements in comparison with the oscillation form of the intact model at the defect value $\bar{t} = 0.75$ is equal to: $\Delta\bar{U} = 55.6\%$ for the first mode of oscillations; $\Delta\bar{U} = 22.3\%$ for the second mode of oscillations; $\Delta\bar{U} = 37.6\%$ for the third mode of oscillations; and $\Delta\bar{U} = 1.31\%$ for the fourth mode of oscillations. For the magnitude of the angle α between tangents of the oscillation format the points of defect location along the rod length, the changes in the corresponding coefficients for the same arrangement of the elastic element are equal to: $\Delta\alpha = 15.7\%$ for the first mode of oscillations; $\Delta\alpha = 5\%$ for the second mode of oscillations; $\Delta\alpha = 83.9\%$ for the third

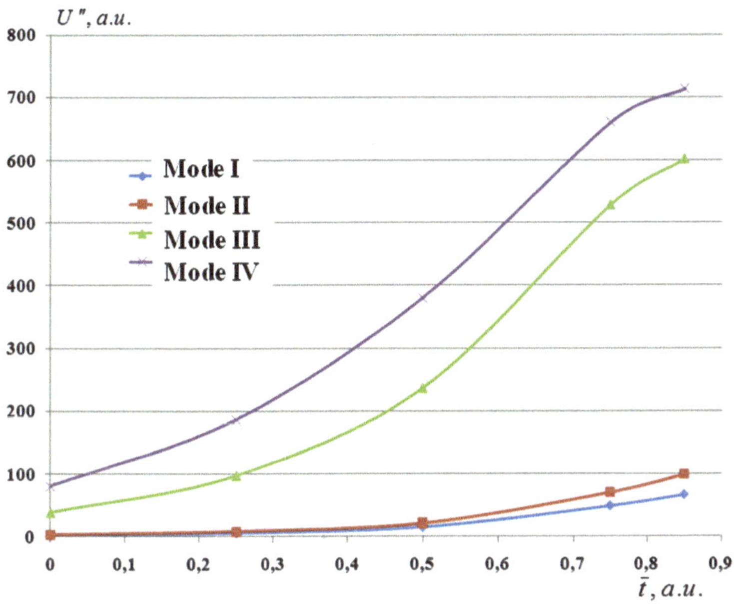

Fig. 6.34 Dependence of the curvature U'' of oscillation form at the defect location site on defect size $\bar{t}$ for different oscillation modes

Table 6.21 Angles α between the tangents on plot of oscillation forms at the point of defect location and their relative values $\Delta\alpha$, depending on the rigidity K_t ($\bar{t}$is the defect size) for different oscillation modes

		Mode I		Mode II		Mode III		Mode IV	
K_t, N·m/rad	$\bar{t}$, a. u.	α,deg.	$\Delta\alpha$,%	α,deg.	$\Delta\alpha$,%	α,deg.	$\Delta\alpha$,%	α,deg.	$\Delta\alpha$,%
1	2	3	4	5	6	7	8	9	10
12,684	0	178.4	0.0	179.5	0.0	169.1	0.0	173.3	0.0
4,531	0.25	176.1	−1.3	178.8	−0.4	149.3	−11.7	163.3	−5.8
1,173	0.50	170.1	−4.7	176.9	−1.4	74.5	−55.9	115.8	−33.2
163	0.75	150.4	−15.7	170.6	−5.0	27.2	−83.9	34.2	−80.3
52	0.85	138.7	−22.3	166.7	−7.1	23.3	−86.2	28.6	−83.5

mode of oscillations; and $\Delta\alpha = 83.5\%$ for the fourth mode of oscillations. For the curvature of oscillation forms at the location $\bar{L}_c = 0.25$ of the elastic element and the magnitude of the defect $\bar{t} = 0.75$ for different modes of oscillation, we have: $\bar{U}'' = 49.3\ \mathrm{m}^{-1}$ for the first mode of oscillations; $\bar{U}'' = 70.3\mathrm{m}^{-1}$ for the second mode of oscillations; $\bar{U}'' = 529\mathrm{m}^{-1}$ for the third mode of oscillations; and $\bar{U}'' = 660\mathrm{m}^{-1}$ for the fourth mode of oscillations.

Analysis of the dependencies of the relative magnitude of transverse displacements on the defect size at the point of the defect location for different oscillation modes (Fig. 6.33) shows that this dependence is well manifested in this defect

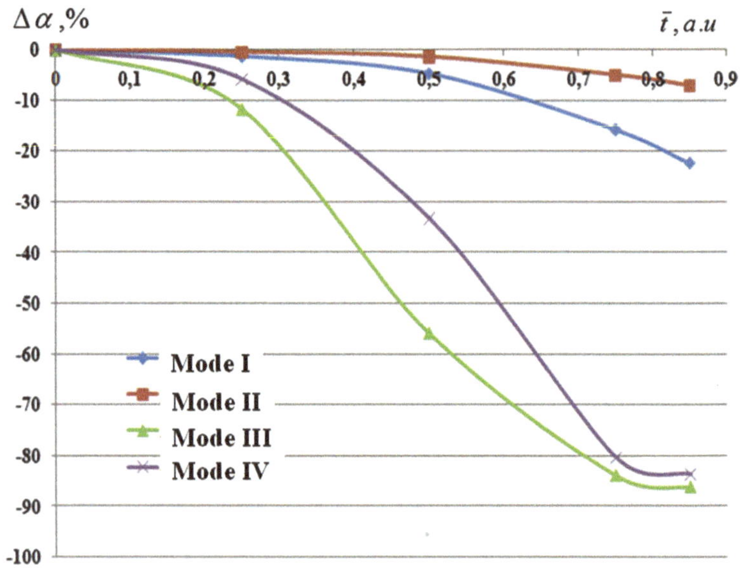

Fig. 6.35 Dependence of the relative magnitude of the change in the angle $\Delta\alpha$ between tangents of oscillation format the point of defect location on the notch size $\bar{t}$ for various oscillation modes

location for the first, second, and third modes and is less clear for the fourth oscillation mode.

Analysis of the dependence of the relative magnitude of the curvature of the oscillation forms at the point of the defect location on the defect size (Fig. 6.34) for different modes shows that all the plots increase monotonically. This dependence for the defect location is well manifested for the first and second modes of oscillations, the third and fourth modes of the oscillations are slightly less sensitive to the notch magnitude.

Analysis of the dependencies of the relative magnitude of the change in the angle of the "break" of the oscillation shape between the tangents at the point of the defect location on the defect size (Fig. 6.35) for different oscillation modes shows that all plots decrease monotonically in different degrees. For the case of defect location, this dependence is well manifested when defect sizes are $\bar{t} > 0.3$ for the third and fourth modes of oscillations, and it is not expressed for the first and second modes.

Analysis of the features of the oscillation modes of the cantilever with damage shows the following:

(i) The plots of oscillation forms have "breaks" at the points of the defect location; in this case, if the site of the defect (elastic element) location is disposed in the zone of bending of the oscillation form curve or in its vicinity, the "break" of the oscillation form curve is poorly identified, and thus this mode will be weakly sensitive to the location of the defect in the structure;
(ii) The parameters of the angle α between the tangents and the curvature $\bar{U}''$ at the points of the oscillation forms are more sensitive compared to the amplitude

parameter $\bar{U}$ of its own oscillation form to the determination of the defect location.

By identifying the defect size, an approach can also be applied based on an analysis of the analytical model and a discrete estimation of the oscillation parameters near the previously determined defect location $\bar{L}_c = 0.25$. So, the stiffness of the rod at the location of the defect can be obtained as

$$K_t = \frac{-EJ\,U_1''(L_c)}{\left[U_1'(L_c) - U_2'(L_c)\right]}. \tag{6.35}$$

The curvature in this case is determined from Eq. (6.31), the angle of rotation of the normal with a discrete approach can be defined as a finite difference:

$$\bar{U}_i' = \frac{\bar{U}_i - \bar{U}_{i-1}}{\Delta x}. \tag{6.36}$$

Let us consider an example of determining the stiffness of a defect by using Eq. (6.35). In the analysis, the case of the location of elastic element $\bar{L}_c = 0.25$ with the stiffness of the elastic element $K_t(0.75) = 163$ N·m/rad was considered. The first and second modes of transverse rod oscillations are investigated by using the analytical model. For comparison, the value of the discrete segment Δx was taken equal to 1/60, 1/100, and 1/240 of part of the cantilever length L.

The deviation in determining the stiffness value for a given case when compared with its original value is calculated as

$$\Delta K = \frac{\left|K_t^* - K_t\right|}{K_t} \times 100\%. \tag{6.37}$$

The results of solving the problem are given in Table 6.22.

Analysis of the stiffness values obtained based on the oscillation shape study shows that for a given defect location $\bar{L}_c$, the deviation of the calculated stiffness using Eq. (6.34) for the first mode of oscillation at discrete interval $\Delta x < L/60$ does not exceed 1.4%. The use of the parameters of the second oscillation mode is possible when the shape of the oscillations is known with great accuracy (at $\Delta x < L/1000$). This is due to the defect being located near the zone of bending the oscillation form.

Table 6.22 Reconstruction of defect rigidity at its location $\bar{L}_c = 0.25$ and discrete interval Δx

	Discrete interval Δx							
	$L/60$		$L/100$		$L/240$		$L/1000$	
Oscillation Mode	K_t^*,(N·m/rad)	ΔK, %	K_t^*,(N·m/rad)	ΔK, %	K_t^*,(N·m/rad)	ΔK, %	K_t^*,(N·m/rad)	ΔK, %
1	165.3	1.4	164.4	0.9	163.5	0.3	163.1	0.1
2	−34.2	–	44.8	–	113.8	30.2	151.2	7.2

6.5.4 Algorithm of the Method for Identifying the Parameters of Defects in Cantilever

Based on previous studies, an algorithm was developed for diagnosing the location of a defect by using the calculated curvatures (see Fig. 6.36). Moreover, it is present in a structural diagram (Fig. 6.37) of the method for identifying the location of a defect in a rod structure under the experimental approach.

At the first stage, information is collected on the natural frequencies and the corresponding oscillation modes of the rod construction. For this purpose, the actual model, oscillation control devices, and computer-based data collection are prepared. With the help of the control unit, the harmonic oscillations of the construction are excited. The oscillation parameters are collected using sensors in several points of the model. The result is the amplitude-frequency response (AFR) of the construction at some points $U_i(x_k)$, then the resonance frequencies are determined, and the data are saved.

Next, information is collected on the modes of natural oscillations at the selected resonance frequencies. With the help of the oscillation control unit, oscillations are excited at the corresponding resonance frequency ω_{ri}. The amplitudes of oscillations are measured at different points along the length of the structure. By combining these data into the array, we obtain the shape of the construction's oscillations at k-ths points $U_j(x_k, \omega_{ri})$, $V_j(x_k, \omega_{ri})$, $W_j(x_k, \omega_{ri})$ at the corresponding resonance frequencies ω_{ri}. The angles $\phi_{ri}(x_k, \omega_{ri})$ between tangents together with curvatures $U''_{ri}(x_k, \omega_{ri})$ are calculated at the corresponding collected points for the amplitudes of the resonance oscillation forms. The probable location of the defect is determined on the base of an analysis of the parameters of modes and the detection of "breaks" on them. Then the data on the parameters of the oscillation forms are saved.

The problem of determining the defect size is solved. Corresponding finite element or analytical model of a construction with a defect localized in a previously defined site is created. The natural oscillations of the rod construction with different sizes of the defect are modeled. The following dependencies are determined: (i) the angles of bending between tangents versus defect size $\bar{t}$ and (ii) the curvatures at the location point of the defect vs defect size $\bar{t}$. Based on a comparison of the results of the experiment and the obtained dependencies of the parameters of the oscillation forms, the defect size is determined.

At the final stage, the adequacy of the calculated and experimental models is evaluated by comparing the natural frequencies and forms of oscillations. The results of the research are the calculated values of location $\bar{L}_c$ and depth $\bar{t}$ of the defect.

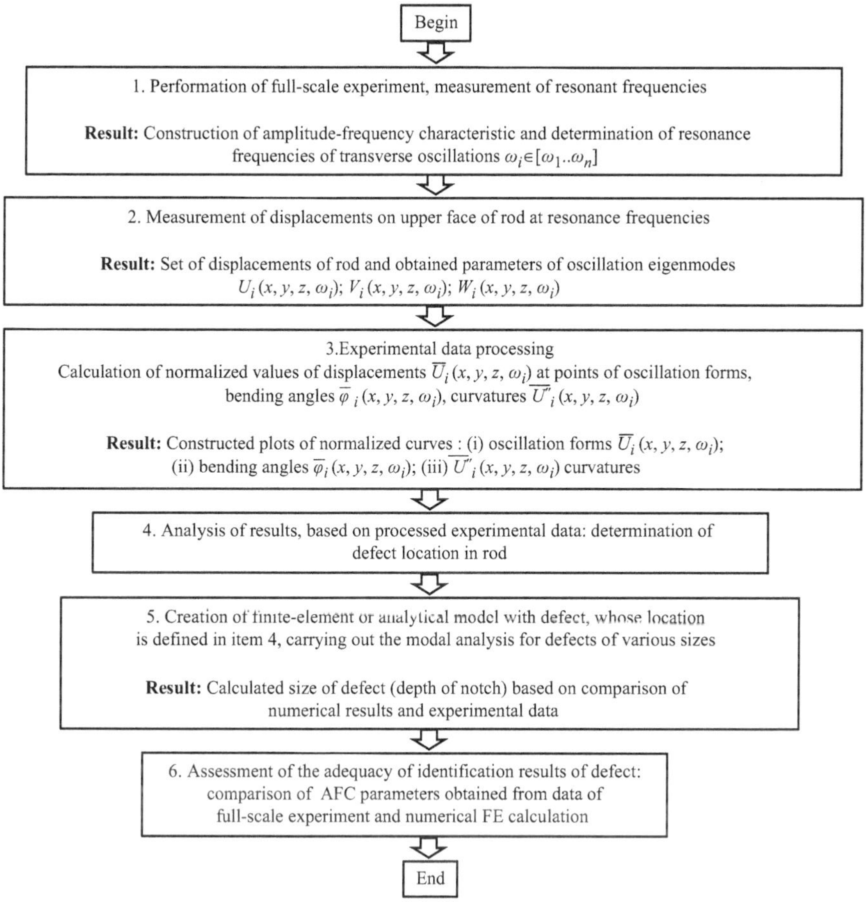

Fig. 6.36 Algorithm for identifying the defect location and size based on the method of calculating the curvature of the oscillation shape of rod element

6.5.5 *Identification of Defects in Rods with Different Variants of Fixing*

The problem of identification of defects in rods with different variants of fixing is considered on the basis of the method of multi-parametric identification in the analysis of frequencies and parameters of the forms of natural oscillations of the construction. We consider rod constructions with one and two defects. The calculation of natural oscillations of the rod is modeled by using the finite-element software ANSYS. In this study, we identified the defect location in the rod, and also compared the parameters of the rod models that had different variants of fixing.

The object of the simulation is a rod (length $L = 250$ mm, height of the cross-section $h = 8$ mm, width $a = 4$ mm) with defects (defect in the form of a transverse

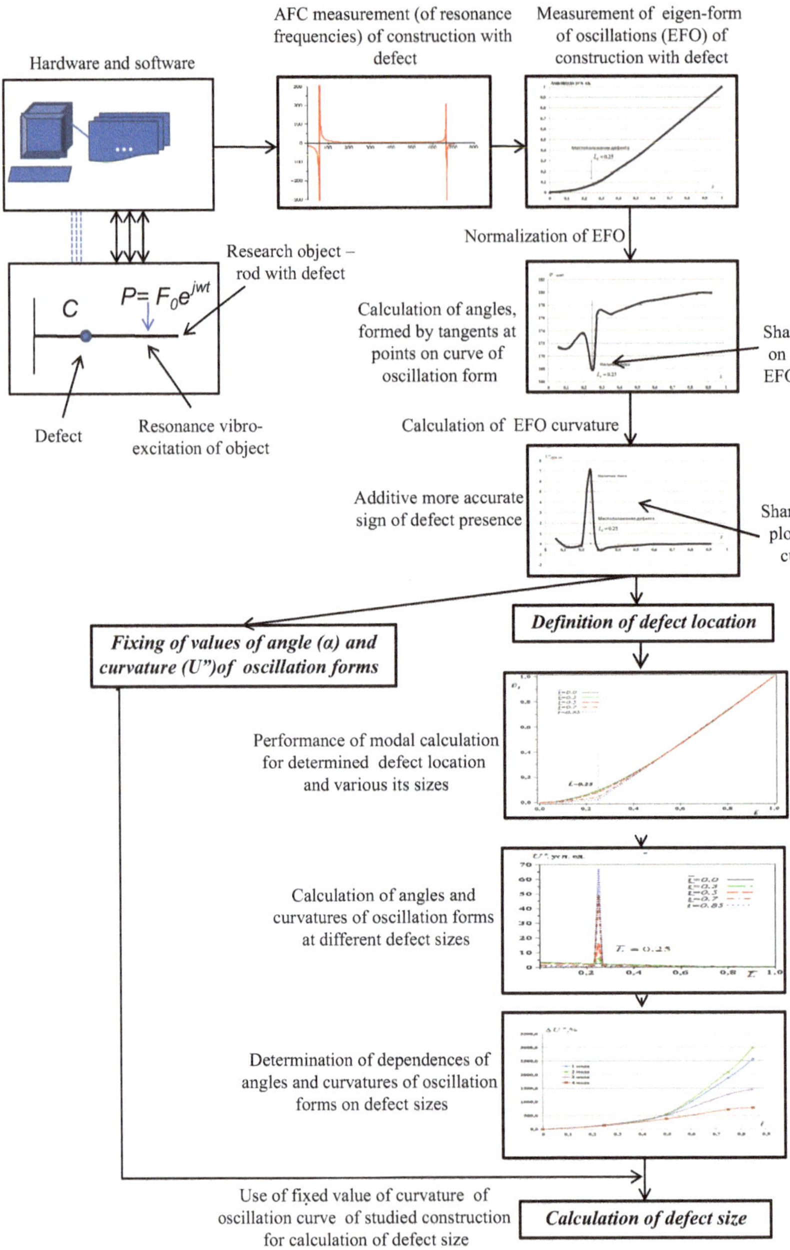

Fig. 6.37 Structural scheme of the identification method for defect location and size in rod construction

Table 6.23 Variants of modeling of rods with different arrangement of defects

No	Fixation variant	Quantity of defects	Location of first defect, $\bar{L}_d$	Size of first defect, $\bar{t}$	Location of second defect, $\bar{L}_d$	Size of second defect, $\bar{t}$
1	(i)	1	0.25	0.7	–	–
2	(i)	2	0.25	0.3	0.7	0.7
3	(ii)	2	0.25	0.3	0.7	0.7
4	(ii)	2	0.25	0.7	0.7	0.7

notch 1 mm wide and depth h_d), located at the point of the rod disposed from the pinch at a distance $\bar{L}_d$, where $\bar{L}_d = L_d/L$ is the location of the notch. We investigate rod constructions with one and two defects. We consider the rods that have two variants of fixations: (i) one edge of the rod is fixed and (ii) the displacements of two edges are fixed. Next, we introduce a dimensionless coordinate $\bar{x} = x/L$, relative depth of damage $\bar{t} = h_d/h$, and consider transverse oscillations of the rod (Table 6.23).

The oscillation simulation is performed by using finite-element software ANSYS. Figure 6.38 presents the finite-element models under study. A partition of the model was chosen and separated into nodes, disposed along the length with a factor of 1/40 of the rod length. The height and width of the rod have a partition into nodes with a factor of 1/3 of the corresponding face. The defect in the form of a notch, reflecting the full-scale model, had a width of 1 mm perpendicular to the cross-section. The finite element mesh had a double concentration near the defect. At the same time, the total number of finite elements exceeded 5,000.

The problem of the natural oscillations of the rod was solved. Forms of the rod oscillations are obtained for different sizes of the defect. Figure 6.39 presents the normalized eigen-modes of the rod vibrations for different sizes of the defect.

The first form of oscillations is considered. An analysis of the oscillation form plots shows that there is a characteristically pronounced "break" at the defect site, depending to a varying extent on the defect size. The break in the curve of the oscillation shape can be weakly manifested, as can be seen in the plots. As an indicative characteristic of the presence of a defect in the cantilever rod, the angle α at the point, formed by the tangents to the curve of the oscillation form, and the curvature of the oscillation form curve are used.

The length of the discrete segment Δx in calculating the parameters of the shape of the oscillations was taken equal to 1/100 part of the rod length. The results of the calculations are shown in Fig. 6.39.

Analysis of oscillation forms (OF), plots of angles at points, formed by tangents, and curvature shows that the use of the criterion for identifying the locations of defects is possible when analyzing these parameters. For the cantilever rod, two parameters well determine the defect location. Due to the angles at the points formed by the tangents being more sensitive to bending the oscillation shape than for a rod with fixation from two edges, the location of the defects is identified weakly. For this variant of fixation and analysis of the shape of oscillations, the defect at its size $\bar{t} > 0.5$ is identified sufficiently well. For the model of rod No. 3, the defect with the

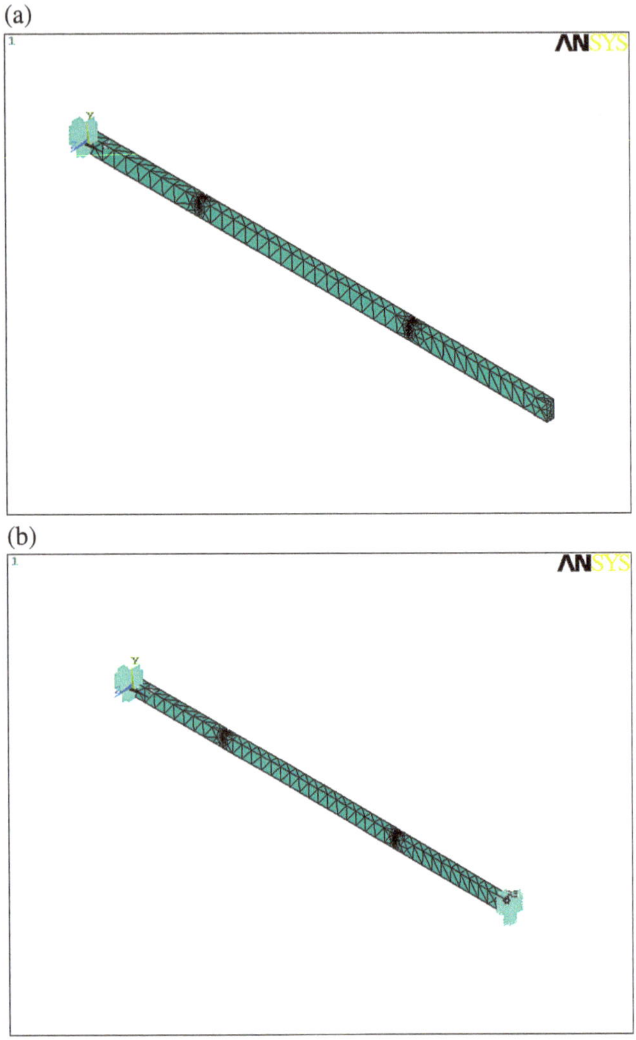

Fig. 6.38 Finite element models of rod with one or two defects: (**a**) cantilever-fixed rod; (**b**) rod rigidly fixed at the edges

location $\bar{L}_d = 0.25$ is poorly identified because it locates at the point of bending the oscillation form. The curvature of the oscillation shape for the case of rod model No. 4 identifies the location of defect well.

The research shows that application of the described method of multiparametric identification of defects in the rod construction makes it possible to calculate the defect parameters in rods having different boundary conditions, including depth and location of the defect. Reduction of the error in the defined parameters of the rod identification is achieved due to the use of a wider set of initial data in the

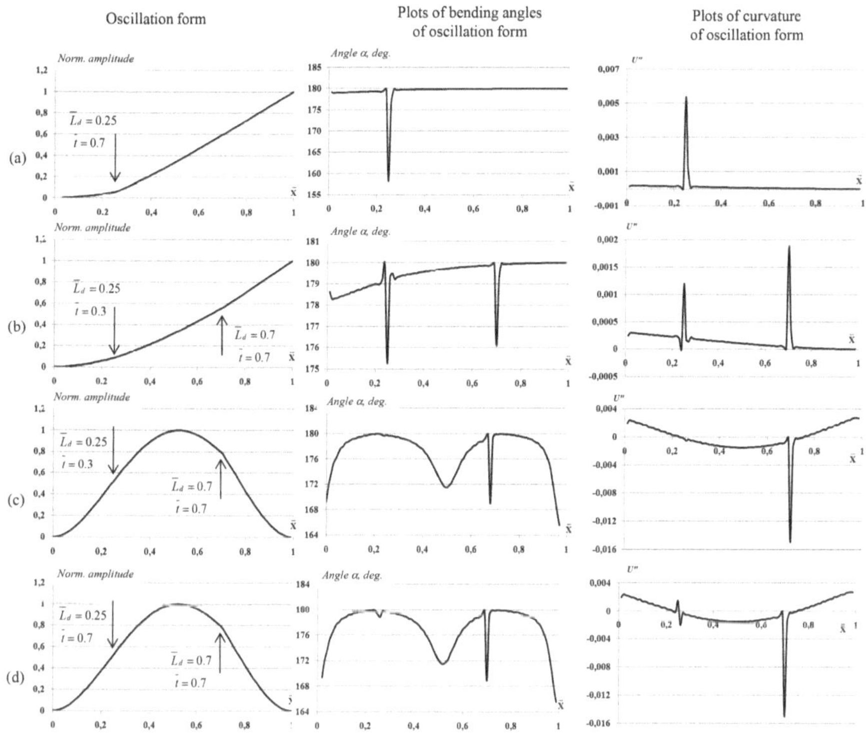

Fig. 6.39 Normalized values of the transverse displacements $\bar{U}$ of rod with defects of different sizes $\bar{t}$ for various variants of its fixing for first mode of transverse oscillations, curves of angles of bending and curvature of first oscillation form

algorithm, and also because of the use of the multiparametric diagnostic sign of identification in the algorithm. The considered method can be used as the base for the development of a method of technical diagnostics of a technical condition of building constructions.

6.6 Conclusions

1. The finite-element calculation of the modal parameters of full-body models of a cantilever rod with defect by using the finite-element software ANSYS was performed. The oscillation forms of the model were presented. The dependencies of natural frequencies on the defect location and size were investigated. The most sensitive modes of oscillations were determined within the dependency on the defect size at its different locations.

2. The analytical model of transverse oscillations of the cantilever elastic rod with defect was considered in the framework of the Euler-Bernoulli model. Dependencies of resonance frequencies on the location and rigidity of the elastic element were obtained. The analysis was performed on the change in the natural frequencies on the location and rigidity of the elastic element for analytical model.
3. The calculation of the dependency between the defect (notch) size of the cantilever rod of full-body finite element model and the flexural rigidity of the elastic element was performed for the analytical model on the base of the dynamic equivalence of models.
4. A comparison was made of the oscillation modes of the first four modes for analytical and FE models of cantilever with different defect sizes in its disposition at the same place.
5. It was shown that the features found in the form of "breaks" and local extremes of the angle α between tangents and of the curvature U'' on the forms of different modes of bending oscillations, coinciding with the location of the defect in the cantilever, can serve as one of the diagnostic features of defect identification and allow the determination of a defect location.
6. It was shown that the angle α between the tangents and the curvature of the oscillation forms of the first four modes at the location point of the defect can serve as a diagnostic sign of identifying its size.

Chapter 7
Set-up for Studying Oscillation Parameters and Identification of Defects in Rod Constructions

7.1 Technical Diagnostics of Defects in Rod Constructions

Chapter 6 considered examples of rod constructions with defects, as well as algorithms for identifying the parameters of the defects. For early identification of the defect, based on the presented approaches, it is necessary to use automated diagnostic measuring systems. At the same time, technical diagnostics of complex systems (e.g., industrial objects and facilities monitoring their condition) are impossible without hardware, software, and methodological support.

Therefore, the main purpose of this chapter is the development of a measuring set-up that allows for the technical diagnostics of rod constructions in practice. The set-up is based on the principle of recording the parameters of oscillations, which allows one to evaluate the parameters of construction defects. In the second stage, the damaged state is determined on the basis of the methods developed. Finally, we present the developed algorithms and the software and laboratory set-up that are needed for the process of identifying defects in rod constructions.

7.2 Measuring Set-up for Identification of Defects in Rod Constructions

7.2.1 Technical Capabilities of the Set-up

The set-up is a test multichannel multiparametric information-measuring system, consisting of three parts:

(i) The electronic hardware for receiving, scaling, converting, and transmitting signals from primary recorders-converters of vibration parameters of the construction;

S. N. Shevtsov et al., *Piezoelectric Actuators and Generators for Energy Harvesting*, Innovation and Discovery in Russian Science and Engineering,
https://doi.org/10.1007/978-3-319-75629-5_7

(ii) The non-conventional dynamic, mechanical, and electromagnetic loading devices for excitation of natural and forced oscillation loadings on the object under study;

(iii) The software "PowerGraf", developed by the company L-card, and original computer program software "Vibrograf." The software is designed for recording, processing, and storage of analog and discrete electrical signals recorded with the help of analog-to-digital converters (ADCs), and allowing a personal computer to act as a standard recording device. The software "Vibrograf" is also designed to control the process of vibration excitation of the object, which makes it possible to automate the process of vibrodiagnostics.

This set-up allows one to conduct the following technical operations:

1. *Registration of oscillation parameters*. With this aim, optical displacement transducers (RF-603) are used for non-contact measurement of the amplitudes of vertical and horizontal oscillations. For high-frequency oscillations, a contactless optical interference transducer of the reflective type (OIT-204) is used to demonstrate an increased frequency response. This transducer is designed and manufactured by E. Rozhkov in the Institute of Mathematics, Mechanics and Computer Sciences, Southern Federal University. Measurement of vertical and horizontal vibration accelerations of oscillations at various points of the rod is carried out with the help of vibration sensors ADXL-103 and ADXL-203. Measurement of deformations on the rod surface is performed with the help of strain gauges of electric resistance TR (with a base of 5 mm). Measurement of deflection is carried out with the help of an optical meter of microdisplacements (OMM). The SU-210 matching device is used to power the OMM sensors. This matching device is able to conduct the primary processing of electrical signals, their scaling and matching on electrical resistance by using input and output plug-in parts.
2. *Excitation of oscillations*. The set-up allows dynamic testing of various rod constructions taking into account both natural and forced oscillations. The natural damping oscillations are excited by the impact method at various points by means of a shock hammer, a ball, or a pulsed electric drummer. The forced oscillations are excited at various points of the rod with the help of an electromagnetic exciter.
3. *Dynamic data processing*. The subsequent processing of signals is performed by a program using the external module ADC/DAC E14-440, developed by firm L-Card, a personal computer and corresponding software. For receiving and processing signals, the software "PowerGraf" is used, developed by L-card. The original computer program software "Vibrograf" is used to solve specific research tasks. With the help of the "Vibrograf" software, the amplitude-time characteristics (ATCs) of the signal are performed, as well as processing and construction of spectrum and amplitude-frequency characteristics of the oscillations. Moreover, it is possible to obtain a form for various modes of vertical and horizontal oscillations and to determine the attenuation coefficients of free

oscillations, and to control vibroexcitation of the oscillations of construction. Observation of the form of the exciting signal from the sensors is controlled by means of a digital oscillograph. Accurate measurement of the signal frequency is performed by the frequency meters Ch3–33 and SFG-2014. Adjustment and calibration of the receiving sensors are carried out using a measuring microscope. Amplification of the electrical signal for the exciter is performed with the help of a power amplifier. The test results are stored in a digital format and a hard-copy (i.e., printed) format.

7.2.2 Development of Structural Parameters of the Set-up

The measuring set-up consists of devices for static and dynamic loading of rod models, a module for calibrating the transverse displacements of the test model, primary sensors, and electronic equipment for recording and processing signals. The measuring set-up is shown in Fig. 7.1.

The static loading device (point 4 on Fig. 7.1) is intended to bend the tested construction element model in the calibration mode of the optical sensors (19a and 19b) and activates the acoustic emission (AE) signals. The loading unit (4) is mounted on the base (15) of the set-up. To control the process of static loading (used to calibrate sensors in the set-up) there is an electronic unit (3), which allows one to carry out the loading process in both manual and program modes using a computer (12).

Dynamic loading of the model at a preliminary given frequency is carried out with the help of an electromagnetic oscillator (17) mounted on the base (15). The oscillator (17) is powered by a low-frequency generator (18). The oscillation frequencies of the model are recorded by a frequency counter (14). It is also possible to excite oscillations through a computer and a DAC.

The module for calibrating and measuring dynamic displacements of the end face of the model includes a measuring microscope (1) fixed in the bracket (2). The control unit of the loading device (3) is mounted on the same bracket. Dynamic displacements of the lower side of the rod (triangular or rectangular configuration) are also recorded by the OIT sensors (19a and 19b) and the vibration sensor (5) (accelerometer of ADXL-103 type). The vibration displacement sensor (accelerometer) (5) and the bimorph piezoelectric actuator (ACT) are located on the horizontal side of the test model. The actuator can be used to excite the loading force on the marked area of the construction rod for damping the oscillations. The outputs of sensors (4, 5, 19a, 19b) and strain gauges (SGs) are connected to the matching device (10) and further to the input of ADC E14-440. Here the signals are digitized and fed to the computer (12).

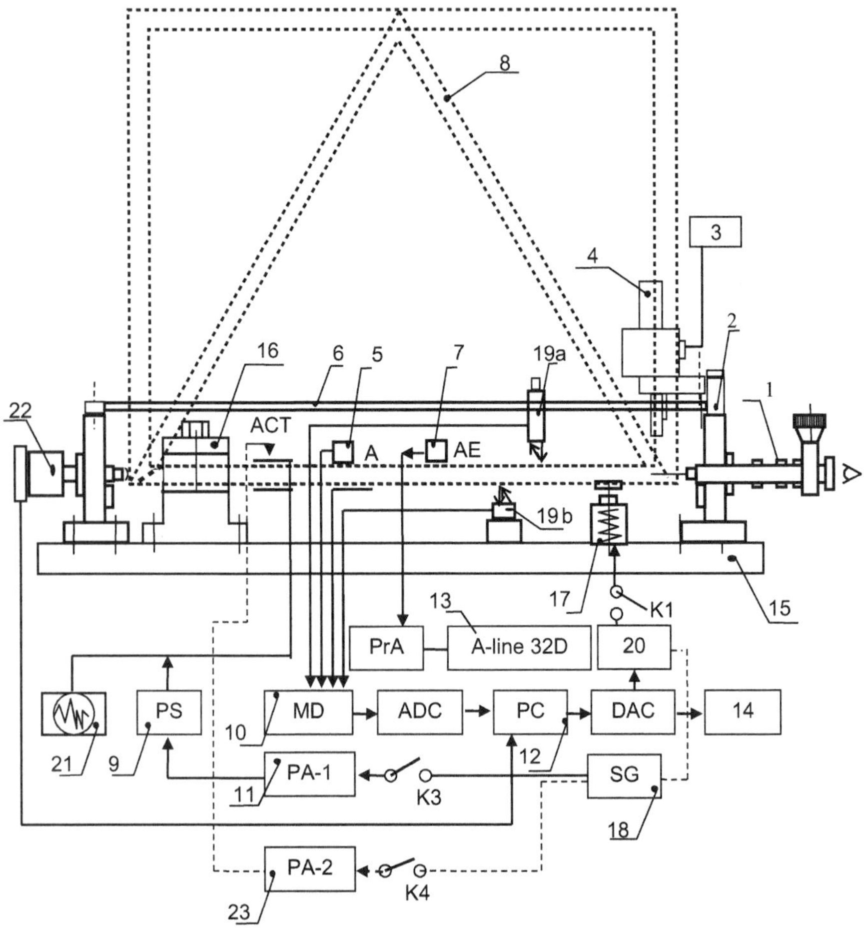

Fig. 7.1 Structural scheme of measuring set-up: 1 – microscope for recording and monitoring the dynamic displacements of test sample (here triangle rod construction); 2 – arm of the microscope; 3 – control unit of loading device; 4 – loading device; 5 – accelerometer; 6 – guides for the OIT sensor; 7 – AE sensor; 8 – sample of the tested model; 9 – phase shifter; 10 – matching device; 11 – power amplifier PA-1; 12 – computer; 13 – acoustic recording system; 14 – frequency counter SFG-2104; 15 – base of the loading unit; 16 – bracket-holder of the test sample; 17 – electromagnetic oscillator; 18 – low-frequency sound generator; 19a –OIT in the vertical direction RF-603; 19b – OIT in the horizontal direction RF-603; 20 – power amplifier; 21 – digital oscilloscope LeCroy; 22 – digital microscope; 23 – power amplifier PA-2; PrA – preliminary amplifier; SG – strain gauges; ACT – piezoelectric actuators; K1, K3, K4 – contactors

7.2.3 *Algorithm for Multiparametric Identification of Defects in Rod Constructions*

In this section, we discuss the algorithm for multiparametric identification of defect parameters in a rod construction with subsequent graphical visualization of resonance frequencies, definition of parameters of natural oscillation modes, location of defects, and their sizes.

In considering the organization of the process of defect identification in rod construction using the measuring set-up and original software, the procedure can be described as being made up of the following steps:

(i) Assembly of the hardware of the diagnostic system, including construction, sensors, amplification device, transmitting path, device for collecting, and processing information;
(ii) Excitation of oscillations and collection of the information on the natural frequencies and oscillation modes of the rod construction;
(iii) Analysis of defects, based on the methods developed in this book.

The result includes the coordinates of the possible location and sizes of the defect. This algorithm is implemented as part of vibrodiagnostics for multiparametric identification of defects in rod constructions.

Figure 7.2 presents the algorithm for multiparametric identification of defects in rod constructions. The algorithm contains the following units:

(i) *Control unit of oscillations* (including computer, DAC, amplifier, vibro-exciter)—controlling the parameters of the forced oscillations of a model;
(ii) *Data collection unit* (including computer, ADC, amplifier, external sensors, and devices for collecting parametric data on the oscillating processes of the construction)—measuring parameters of oscillations of a model;
(iii) *Processing unit* (including computer, applied program modules and representing software tools)—performing the primary processing of the measured signal of the model oscillations;
(iv) *Data bank of rod construction* (including software tools) —collecting and storing information on the modal parameters of the rod construction;
(v) *Analysis unit* (including software tools) —processing of parametric data of oscillatory processes of a model;
(vi) *Information output unit*—allowing one to display graphic data on the parameters of oscillations and to save the report data.

At the first stage (the experimental full-scale model), oscillation control devices and data collection are prepared using a computer with the appropriate software allowing one to control the information collection process, and also to control the vibration parameters and external electronic units of the matching devices.

At the stage of the cyclic process of collecting the primary information on the amplitude-frequency characteristics of the rod construction, the frequency parameters are set for the vibration excitation in the limits $\omega_i \in [\omega_1, \omega_{nw}]$, required for the tests, with their change with step $d\omega = (\omega_{nw} - \omega_1)/(nw - 1)$, where nw is the number of frequencies analyzed. At each step, excitation of the oscillations of the rod construction occurs at frequency ω_l at some point x_k of the construction with the help of the control unit of oscillations. After stabilization of the oscillatory process, a transition take place to the data collection unit. In this unit, oscillation indicators are collected using sensors (e.g., laser triangulation transducers of displacements) at several points.

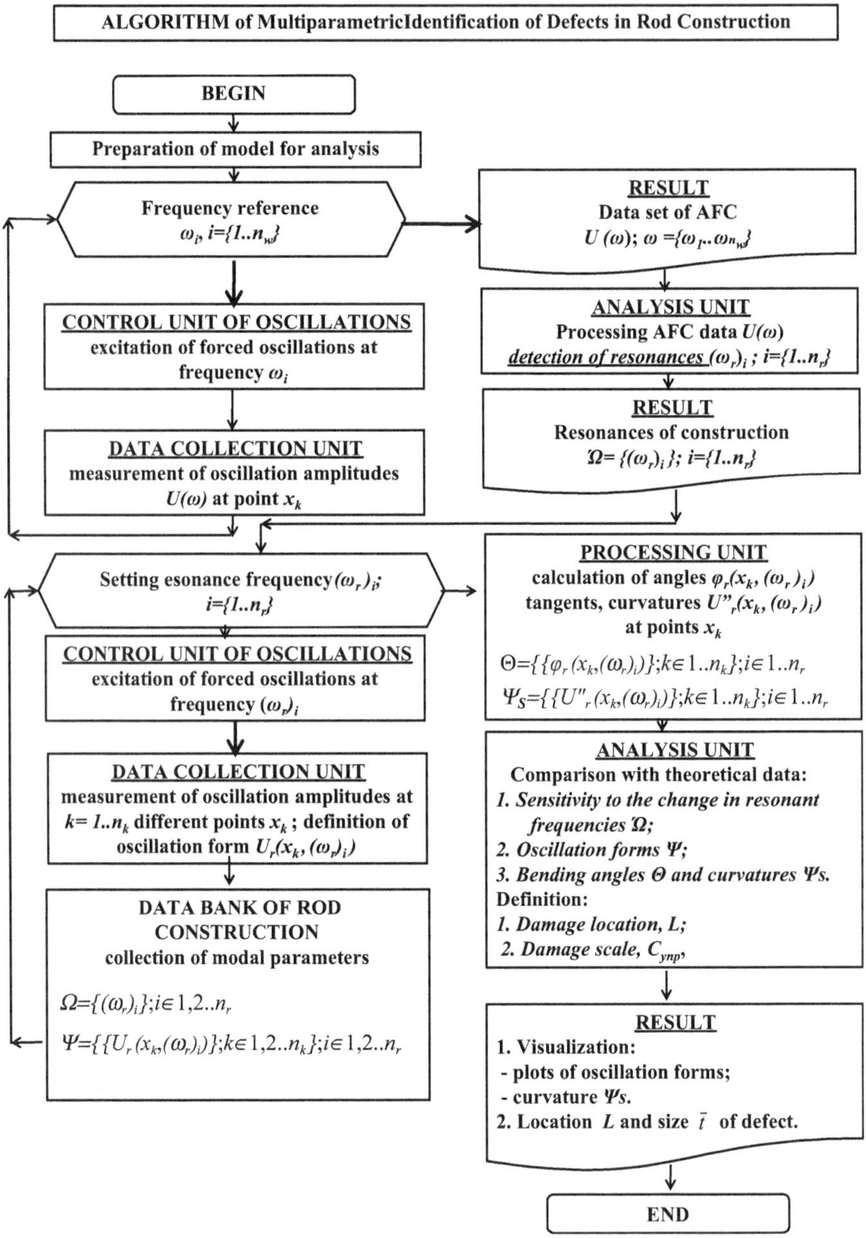

Fig. 7.2 Algorithm of multiparametric identification of defects in rod construction

Upon completion of the data collection, the transition to the beginning of the cycle occurs, the setting of a new frequency $\omega_{i+1} = \omega_i + d\omega$, the repetition of the vibro-excitation of the rod construction, and the collection of information on the oscillations. The result of the operation of the units in the repetition cycle of resonance frequencies ω_i is the data array $U_i(x_k, \omega_i)$ at some point x_k of the construction, presenting itself as the amplitude-frequency characteristics (AFCs) of the construction at a given point. At the next stage, the graphic image of the dependency $U_i(x_k, \omega_i)$ is created with storage of data to file. In the data processing unit, the measured AFCs $U_i(x_k, \omega_i)$ are processed and the resonance frequencies ω_{ri} are determined. In the output unit, information on resonance eigen-frequencies is output to a screen or stored in a file.

The next step is to collect information on the forms of natural oscillations at the selected resonance frequencies. First, the cycle parameters are set in the form of eigen-frequencies ω_{ri} having rn repetitions. With the help of the control unit of oscillations, oscillations are excited at the corresponding resonance frequency ω_{ri} with the number ri. In the data collection unit, the amplitudes of the oscillations are measured at different points along the length of the construction. By combining these data into the array, we obtain the shape of the construction oscillations at k points with coordinates x_k at the corresponding resonance frequency ω_{ri}. In the data processing unit, the primary amplitude characteristics of the corresponding oscillation form are recalculated for each point by means of calculating the angles between the tangents at the amplitude combining points, and constructing the arrays of angles $\phi_{ri}(x_k, \omega_{ri})$ and curvatures $U''_{ri}(x_k, \omega_{ri})$. At the next stage, data on the corresponding oscillation form, the angles of the tangential, and resonance frequencies are stored in the database of rod construction.

After the completion of the cyclic procedure for measuring the parameters of the oscillation forms, calculating the angles of the tangents at the points and the curvature parameters, a transition is performed to the analysis unit. This determines the sensitivity of frequencies to the presence of a defect, analyzes the features of the oscillation forms to determine the location L of the defect, its relative stiffness C_{el} and depth t. In the next output unit, the oscillation forms $U_{ri}(x_k, \omega_{ri})$, angles $\phi_{ri}(x_k, \omega_{ri})$, curvatures $U''_{ri}(x_k, \omega_{ri})$ at points, probable places L and depths t are graphically output on the screen together with a report on the work performed, which is saved in the file.

7.2.4 *Technique of Carrying Out Test Measurements of Modal Characteristics of the Beam Construction*

In this section, we describe the technique of performing measurements. The oscillation parameters of the vibro-exciting of model sample and the collection of measured oscillation parameters are controlled with the help of the software "Vibrograf." The oscillations are excited by means of an electromagnetic vibrator in the measuring set-up, which is installed near the previously selected points of excitation of the oscillations in the beam construction.

Measurement of displacements and deformations is carried out by a strain gauge, piezoelectric, optical displacement sensors, and accelerometers installed at the marked calculated points of the sides of the cantilever beam with the possibility of serial (parallel) temporal registration and spectral processing of signals. The strain gauges are located as close as possible to the pinching and the vibration sensor is installed with a magnetic holder.

In accordance with the above-mentioned method, the algorithm for performing measurement operations is as follows:

(i) The points are marked out, and the oscillation parameters are measured and the vibration sensor is set at the calculated point on the rod face;
(ii) The forced oscillations are excited in the model at the required frequency;
(iii) The deformations, deflections, and vibration displacements are recorded at the corresponding points of the rod (or at the points of the lower side of the triangular structure) by using the Le-Croy oscilloscope or the ADC computer device with the help of all sensors;
(iv) When the resonance frequency is reached, first the vibro-displacements are registered with a vibration sensor on the magnet by hooking the sensor to each point and recording the vibration displacements after stabilizing the vibration of the construction; then the vibration shifts are registered with the help of a laser triangulation sensor (RF603) on a movable beam, positioning at each marked point;
(v) Transition to the excitation of the next frequency is performed;
(vi) For each sample with the notches t_1, $t_2 \ldots t_n$, steps (ii) – (v) are repeated, the frequency changes from 0 to 2,000 Hz and the frequency of vibro-excitation is automatically changed in intervals of 0.5 Hz;
(vii) The results of measurements of amplitude-frequency and modal characteristics of models of construction elements are recorded in measurement protocols in text and graphic formats.

7.2.5 *Software for Automation of the Measurements of the Oscillation Parameters of Beam Constructions*

To automate the measurement of the oscillation parameters and creation of dynamic deformation images of the investigated beam constructions, the software "VibroGraf" was developed and written in Visual Delphi. The development was carried out in the I. I. Vorovich Institute of Mathematics, Mechanics and Computer Sciences, Southern Federal University.

The computer program includes the following modules:

(i) "Spectroscope"
(ii) "Oscilloscope"
(iii) "Spectrograph"
(iv) "Signal View"
(v) "Tuning."

Description of the software "VibroGraf." The module "Spectroscope" allows one to collect data on the amplitude-time characteristics (ATCs) of the steady-state forced oscillations of the tested beam construction in a selected frequency range. Data on the measured image of oscillations (ATCs) are recorded in the memory buffer, the buffer is analyzed and the range and amplitude of the oscillations are selected at the current frequency. The result can be saved as text data or graphically.

The module "Oscilloscope" allows one to obtain actual amplitude-time characteristics of the construction oscillations at a selected frequency or with damped oscillations. The module displays a graphic image of a specific array of data on the measured parameters (amplitudes) of the oscillations. Moreover, the calculation and output of the amplitude-frequency response for this array are performed. This module is intended for primary and more accurate adjustment of the parameters of the vibration sensors of the construction. Additionally, an array of measured oscillation parameters is stored in computer memory.

The module "Spectrograph" is intended for constructing the amplitude-frequency characteristic of the obtained deformation image (amplitude-time characteristic) for a number of points determined by the computer program based on the fast Fourier transform algorithm.

The module "Signal View" (Visualization) is intended for output and processing the data of an array of deformation images (amplitude-time characteristics) from memory.

The module "Tuning" is intended for tuning the channels and frequencies of the analog-to-digital converter.

The procedure for working with the software "VibroGraf" assumes the following actions:

The computer program is developed with an intuitively clear interface and is intended for a mean-trained user. After starting the program, it is necessary to conduct primary tuning of the ADC module. To do this, go to the "Settings" tab (Fig. 7.3).

In panel (3), select the shape and amplitude of the excitation force signal. Window (4) shows the selected excitation force signal. Panel (1) allows the user to tune the ADC amplifier. To select the number of measurement channels, refer to the drop-down list (2) (their number does not exceed the maximum possible for the module E14–440).

In the "Tuning" module (Fig. 7.4), to the user must calibrate the measuring sensors: set the zero position of the mark when calibrating the amplitudes of

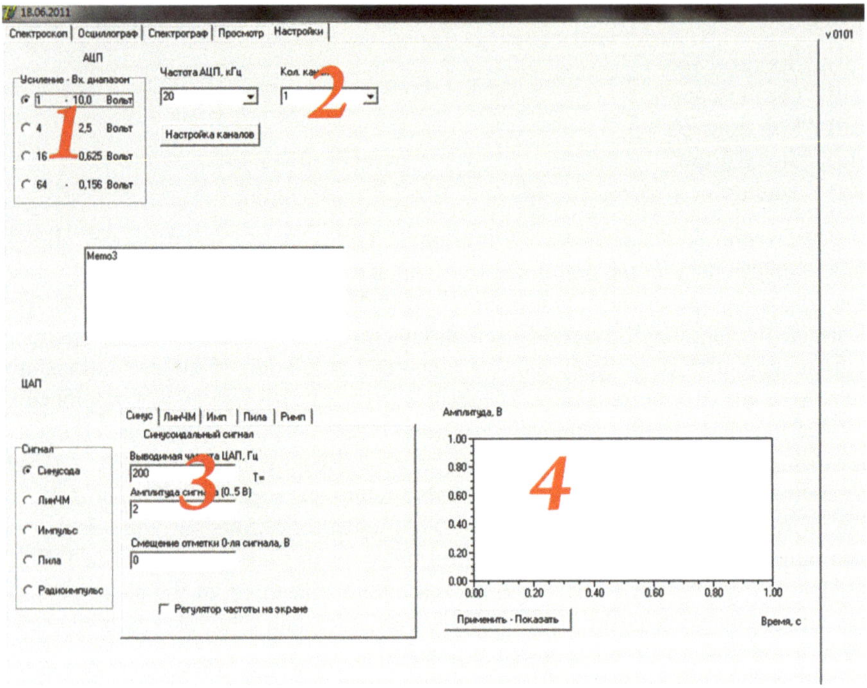

Fig. 7.3 Panel “Settings” of software “Vibrograf”

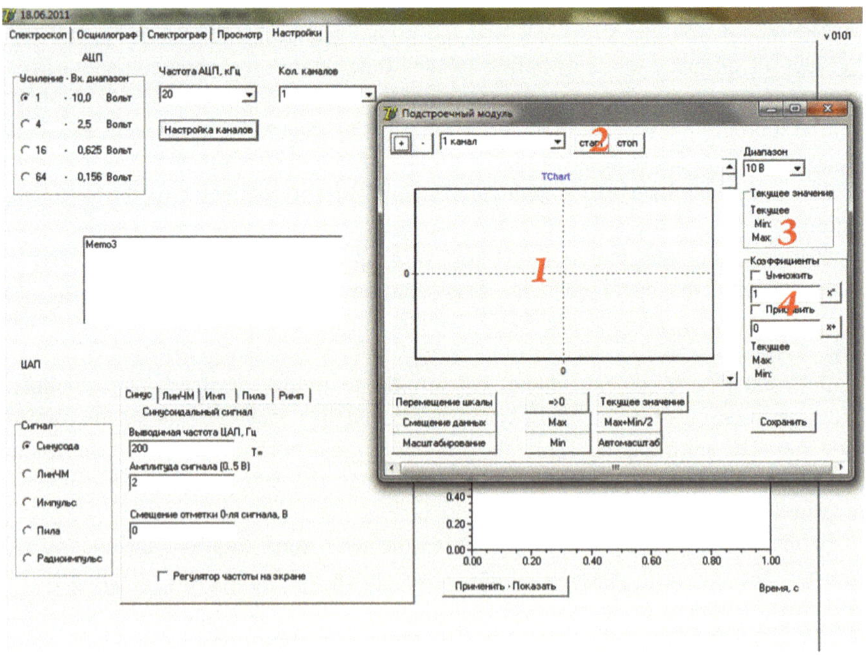

Fig. 7.4 “Tuning” module of software “Vibrograf”

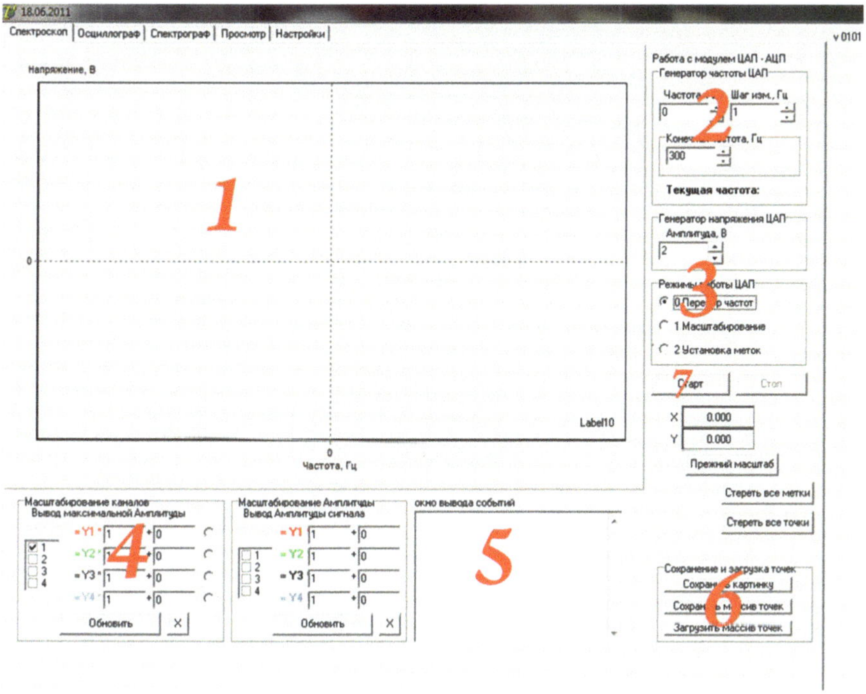

Fig. 7.5 Panel "Spectroscope" of software "Vibrograf"

displacements. Window (1) shows the signal output. First, it is necessary to start the registration of data by using button (2). Panel (3) displays the current maximum and minimum values of amplitude. In (4), by using a constant multiplier and summand, we can tune a convenient signal display.

The organization of the process of collecting the measured ATCs of the construction is performed by sweep method through all frequencies. First of all, we need to adjust the ADC and DAC parameters. In order to organize the process of measuring the ATCs of the construction, we go to the panel "Spectroscope" of the software "Vibrograf" (Fig. 7.5). Panel (2) allows one to set the frequency range in which the study will be performed. Panel (3) allows one to set the voltage amplitude of the DAC, as well as the following modes of operation: frequency scanning, scaling, and labeling.

Panel (4) is needed to scale the signal amplitude and maximum amplitude for different measurement channels. The program reports the current state of the experiment, as well as errors in window (5). It is also possible to save the resulting plot as an image or an array of points, and also load an existing array for plotting. In order to do this, we must refer to panel (6). Finally, when all the parameters are set, button (7) allows one to start the experiment.

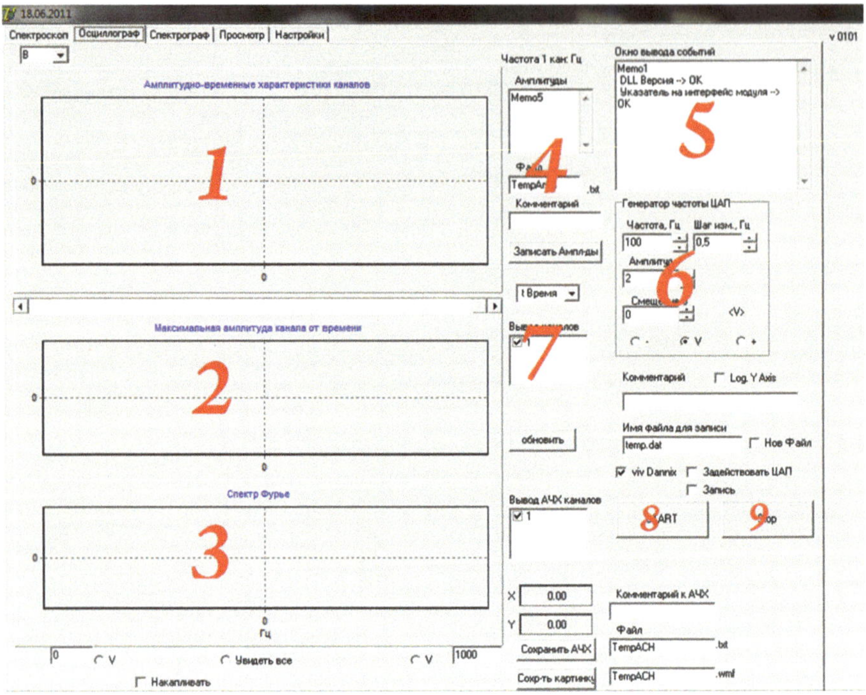

Fig. 7.6 Panel "Oscilloscope" of software "Vibrograf"

First, the program needs to be started in data collection mode, i.e., frequency sampling, after which, when the plot is built, we can determine the maximum (minimum) and set the labels. With this aim, we go to the mode of setting the label in panel (3). Then click on any point of the plot, highlight its coordinates, where the abscissa of a point presents the frequency and the ordinate corresponds to the amplitude. In order to determine the parameters of the vibration excitation signal, we go to the panel "Oscilloscope" of the software "Vibrograf" (Fig. 7.6).

In window (1), after the registration of data, the waveform of the signal is displayed against time. Window (2) displays the maximum amplitude versus time. In window (3), the spectrum of frequencies is displayed using the fast Fourier transform. Panel (4) records the signal amplitude and it is possible to store these data. We can select from the list (7) the measurement channel with which we want to work. First, in order to record the offsets of a particular rod point, we must select the first measurement channel. Window (5) displays the test process and service messages. Panel (6) presents a frequency generator of DAC. Here it is required to set the frequency of the exciting force. Button (8) is needed to start the experiment, and (9) for its stopping.

7.3 Calculation-Experimental Approach to Determination of Defects in Cantilever-Shaped Beam Construction

7.3.1 *Description of Studied Object*

Physical models are represented by beam models. The beam sizes were taken as follows: length $L = 250$ mm, cross-section of rectangular configuration $b \times h = 4 \times 8$ mm^2. Material was St10 (modulus of elasticity $E = 2.068 \times 10^{11}$ MPa, material density $\rho = 7{,}830$ kg/m^3). The left end of the beam had a rigid steel pinch with sizes of $17 \times 28 \times 48$ mm^3. The defect was carried out by cutting a beam 1 mm wide at the site L_{cut}. The depth of the defect in the form of a notch after each AFC measurement was increased by cutting to the values required by the action algorithm compiled earlier.

7.3.2 *Full-Scale Experiment*

An example of measuring the forced oscillations of a beam construction is presented here. The amplitude-frequency characteristics were defined in the frequency range 0–2,000 Hz; this range was chosen from the sensitivity condition of the sensors of the hardware part of the set-up. The distributions of the amplitudes of vertical displacements along the length of the sample were recorded consecutively on the first, second, and third modes of oscillations with the help of moving optical sensors (10) and (11) (Fig. 7.7). According to the obtained data, the oscillation forms of all three investigated oscillation modes were restored. It should be noted that in order to reliably restore the oscillation form, the amplitude of the forced oscillations must be at least an order of magnitude larger than the amplitude of the "noise" caused by mechanical and electrical causes.

The sample of the rod model (1) is mounted on the base (3). The right end of the beam is free, and the left end is rigidly fixed by the bracket of the support holder (2). The steady oscillations of the sample arise due to the harmonically varying lateral concentrated force supplied by the electromagnet (4). The shape and amplitude of the signal are set in the computer software "VibroGraph", and transmitted to the (7) DAC E14–440. Then an electromagnetic oscillator is activated using the preamplifier LV102 (5) exciting oscillations in the construction. The generator G6–27 (6) can serve as a duplicating device for exciting a harmonic load. The frequency counter (8) SFG-2104 and the digital oscilloscope (9) LeCroy WS-422 provide additional control over the exciting frequency and amplitude of the signal from the DAC. Part of the kinetic energy of the beam oscillations is transmitted to sensitive elements (10), (11), and (12). The signal is transmitted via the matching device (13) to the external E14–440 module (7), after which the digitized signal can be reproduced on the computer. Processing the experimental data received in real time from the analog-digital conversion (ADC) module E14–440, was carried out with

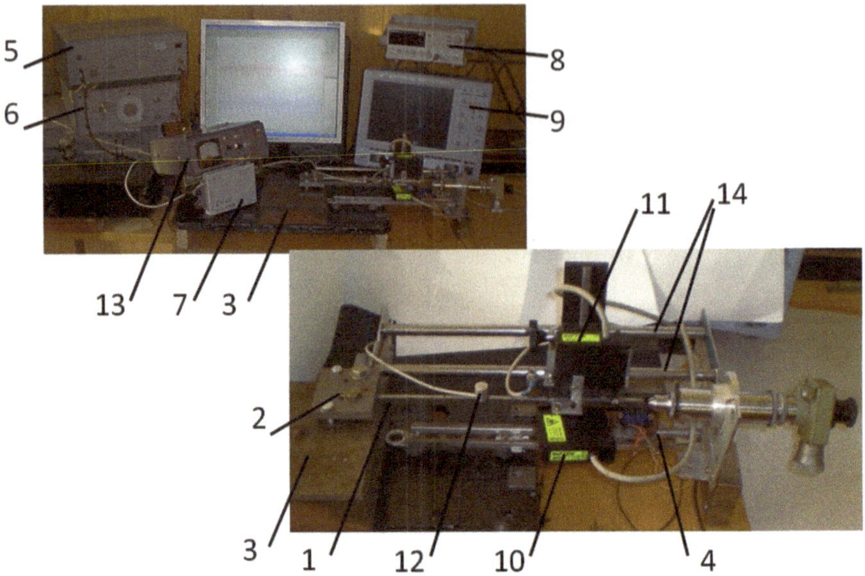

Fig. 7.7 Common view of the measuring set-up: 1 – sample; 2 – support holder; 3 – base; 4 – electromagnetic exciter EMV210; 5 – power amplifier LV102; 6 – generator G6–27; 7 – ADC/DAC E14–440; 8 – frequency counter SFG-2104; 9 – digital oscilloscope LeCroy WS-422; 10 – optical sensor for horizontal measurements RF603; 11 – optical sensor for vertical measurements RF603; 12 – vibration sensor of ADXL-203 model; 13 – matching device; 14 – guiding rods of optical sensors

the help of the software for measuring the amplitude-frequency characteristics ("Vibrograf").

With the help of the guides (14) for the sensor (11), it is possible to change the position of the sensor relative to the test sample, so that deflections can be measured at any point along the horizontal axis of the beam.

7.3.3 Approbation of Calculation-Test Approach for Determination of Cantilever Beam Defects

7.3.3.1 Experimental Studies of Frequencies and Oscillation Forms of Cantilever Beam with Notch Using Measuring Set-up

At the first stage, the AFC was measured at various points of the beam model. The notch was located at the point $L_c = 0.25$. The notch variants were taken as follows: $t = 0.30$ ($a = 2.4$ mm); 0.50 ($a = 4$ mm); 0.70 ($a = 5.6$ mm); 0.86 ($a = 6.9$ mm). A model without notch was also tested ($t = 0$).

Measurements were conducted in the frequency range of 1–2,000 Hz using an accelerometer. The accelerometer (biaxial vibration sensor ADXL-203) was mounted at the point $L = 15$ mm from the pinched end of the beam. Moreover, the measurements were performed of transverse displacements into two *Oxy* and *Oxz* planes using two optical transducers RF-603 of transverse oscillations on the horizontal and vertical faces of the beam.

Examples of the operation of the software "Vibrograf" and "PowerGraf" at measuring ATCs responses during impact and vibration excitations of the beam cantilever model are shown in Figs. 7.8, 7.9, 7.10, and 7.11.

Figure 7.12 shows the frequency response of the beam vibrations at different notches. Measurement of the ATCs of the beam construction with different cut values was carried out by using an accelerometer. The sensor of the accelerometer was located in such a way that its sensitivity was maximal to oscillations in the vertical plane of the rod. Analysis of the amplitude-frequency characteristic shows that resonances, corresponding to the oscillations of the sample in the vertical plane, had the largest response of the amplitude, and a small response of the amplitude related to the resonances of the sample oscillations in the horizontal plane (Fig.7.12).

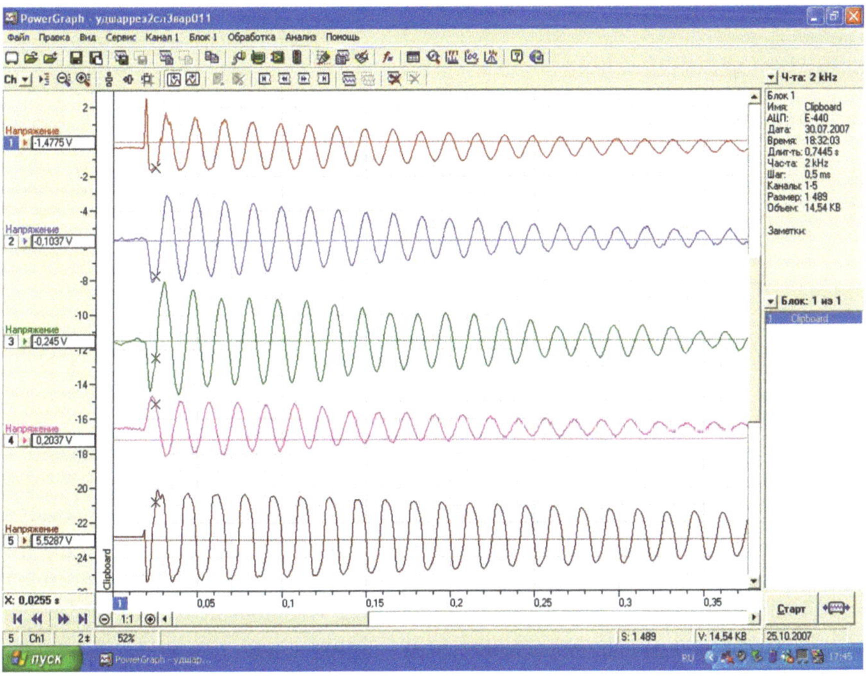

Fig. 7.8 Example of operating software "PowerGraf" at measurement of ATCs of beam damped oscillations by using five sensors

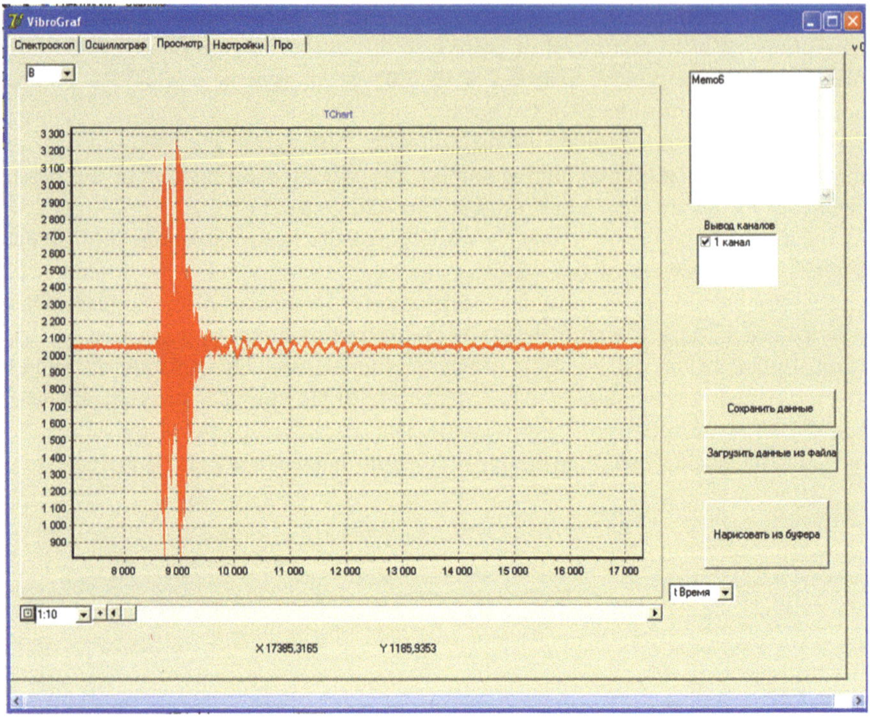

Fig. 7.9 Example of panel "View" of software "Vibrograf": output to working panel of damping signal plot of construction response to impact excitation, measured by accelerometer

7.3.3.2 Finite-Element Modeling

By using finite-element software ANSYS, a three-dimensional cantilever model was developed. The modeling of the beam construction was carried out based on the principles presented in Chap. 6.

In the result of the finite-element calculation of the oscillations of the beam with a notch, a set of natural frequencies and corresponding forms of oscillations were obtained. For comparison, the plots of the beam's own oscillation modes (vertical displacements of the points of the upper face of the beam for the modes of oscillations in the *Oxy*-plane), with a notch located at a point $L_c = 0.25$ and having a different depth t, are shown in Fig. 7.16 a, c, and e. The plots present dimensionless characteristics and parameters, related to the displacement amplitude at a point, corresponding to the free edge of the beam.

The plots show the distribution of the amplitudes of transverse displacements along the beam length L with a range 0.02 L. The calculations were made with a relative depth of the notch $t = t_i/a$ (where t_i is the absolute value of the considered cases of the notch depth, and a is the height of the cross-section of the beam), which assumes values of 0.30, 0.50, 0.70, and 0.86, and for the intact beam $t = 0$.

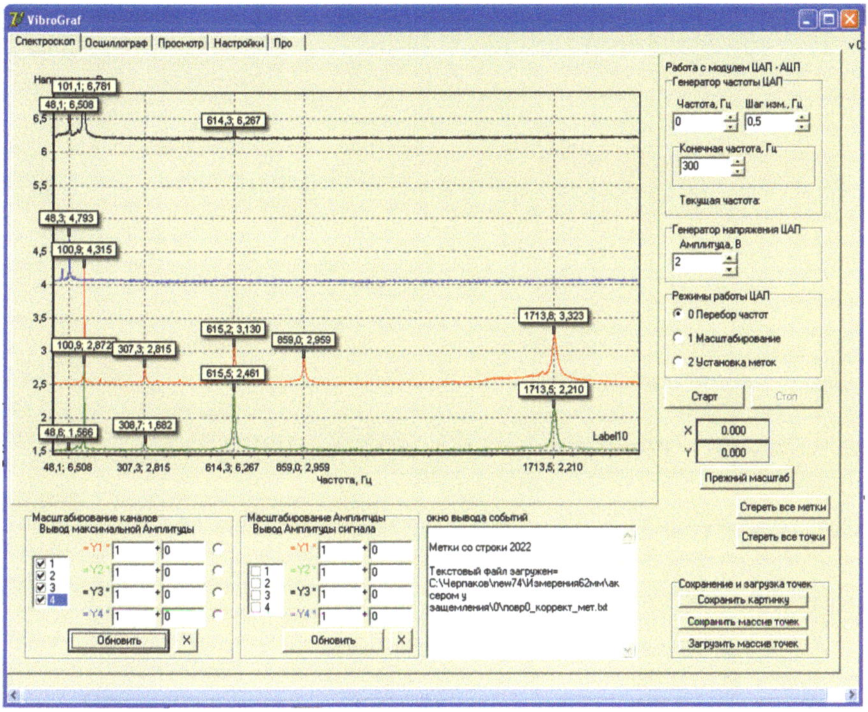

Fig. 7.10 Example of panel "Spectroscope" of software "Vibrograf": definition of beam amplitude response with help of four sensors

The adequacy of the constructed model in the FE modeling in respect to the experimental model was estimated on the base of a comparison of the deviations of the natural frequencies in the simulation of the cases of various degrees of rigidity at the base of pinching the beam. Table 7.1 shows the natural frequencies obtained in the experiments and the FE calculation of the modal parameters of the beam. The calculation of the relative frequency deviation, performed by comparing the frequencies, obtained on the base of experiments and calculation, is given by the following formula:

$$\Delta = \frac{\omega_e - \omega_{\mathrm{FE}}}{\omega_e} \times 100\%. \tag{7.1}$$

The comparative analysis of the natural frequencies of the calculated and experimental models shows that the smallest discrepancy was obtained in the plane of the greatest flexural stiffness of the model, namely the second, fourth, and sixth modes of oscillations. The frequency deviation for the 2–7 oscillation modes is within the range of 12.9%, which is a sufficient criterion for approximation of the model.

At the next stage, the forms of transverse oscillations in the vertical plane of the beam model were obtained. Measurements were performed at 13 points on the upper

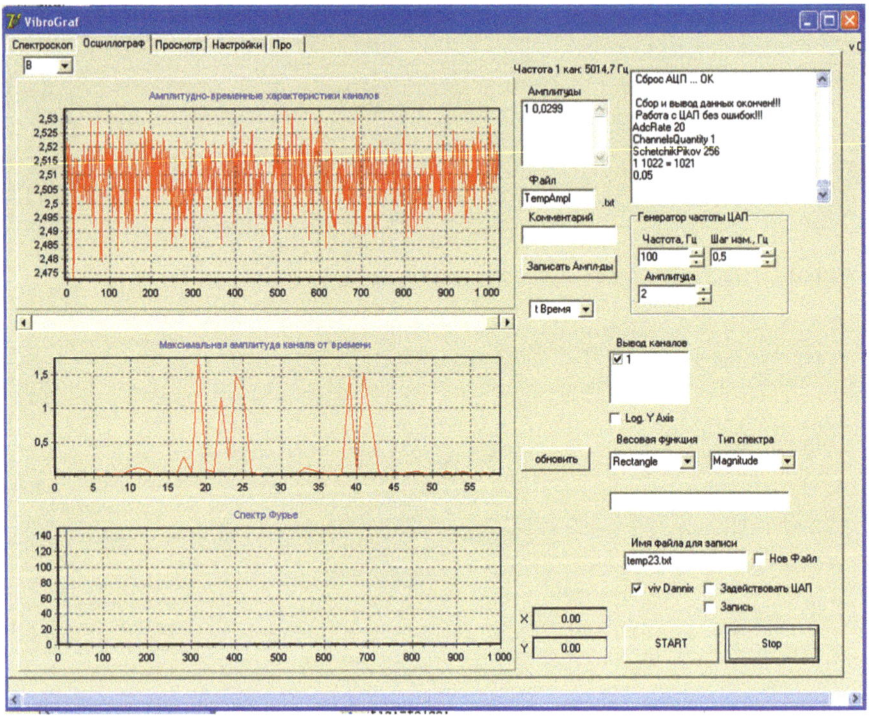

Fig. 7.11 Example of panel "Oscilloscope" of software "Vibrograf": measurement of ATCs performed using a displacement sensor

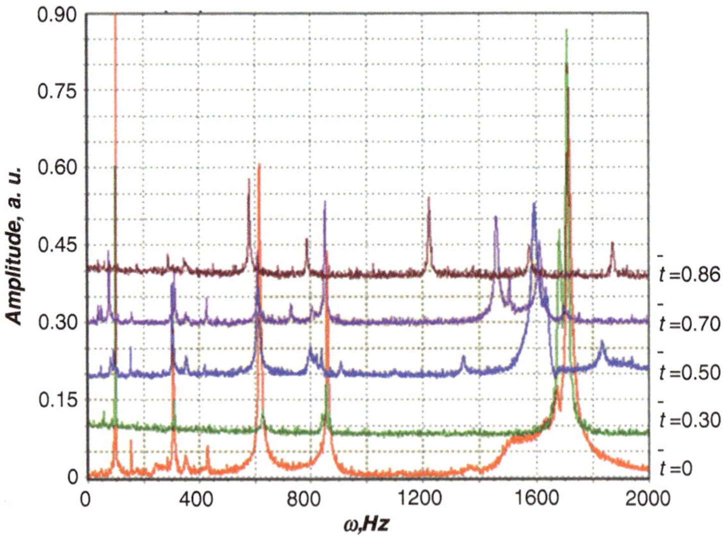

Fig. 7.12 Amplitude-frequency characteristics of beam at distance of $L_c = 0.15$ mm from the pinched end of the beam: measurements performed using accelerometer

Table 7.1 Natural frequencies of beam oscillations, calculated using software ANSYS and obtained in experiments

Number of Oscillation Mode	Normalized size of defect t									
	0		0.30		0.50		0.70		0.86	
	FE	Test	FE	Test	FE	Test	FE	Test	FE	Test
	Deviation Δ, %		Deviation Δ, %		Deviation Δ, %		Deviation Δ, %		Deviation Δ, %	
1	53.0	48.7	52.6	48	52.0	47	50.9	45	42.1	38
	8.1		8.8		9.7		11.6		9.6	
2	97.9	101	95.6	99	90.3	93	75.5	75	48.6	44
	−3.2		−3.6		−3.0		0.6		9.5	
3	332.6	308	332.4	311	332.3	303	331.8	312	330.5	288
	7.4		6.5		8.8		6.0		12.9	
4	621.2	615	618.5	624	613.0	607	598.5	607	574.2	577
	1.0		−0.9		1.0		−1.4		−0.5	
5	931.8	858	927.2	855	919.3	838	904.4	851	874.9	786
	7.9		7.8		8.3		5.9		10.2	
6	1746.4	1675	1704.6	1683	1619.2	1594	1446.8	1455	1258.1	1223
	4.1		1.3		1.6		−0.6		2.8	
7	1826.7	1713	1817.6	1710	1802.9	1637	1776.6	1611	1729.5	1576
	6.2		5.9		9.2		9.3		8.9	

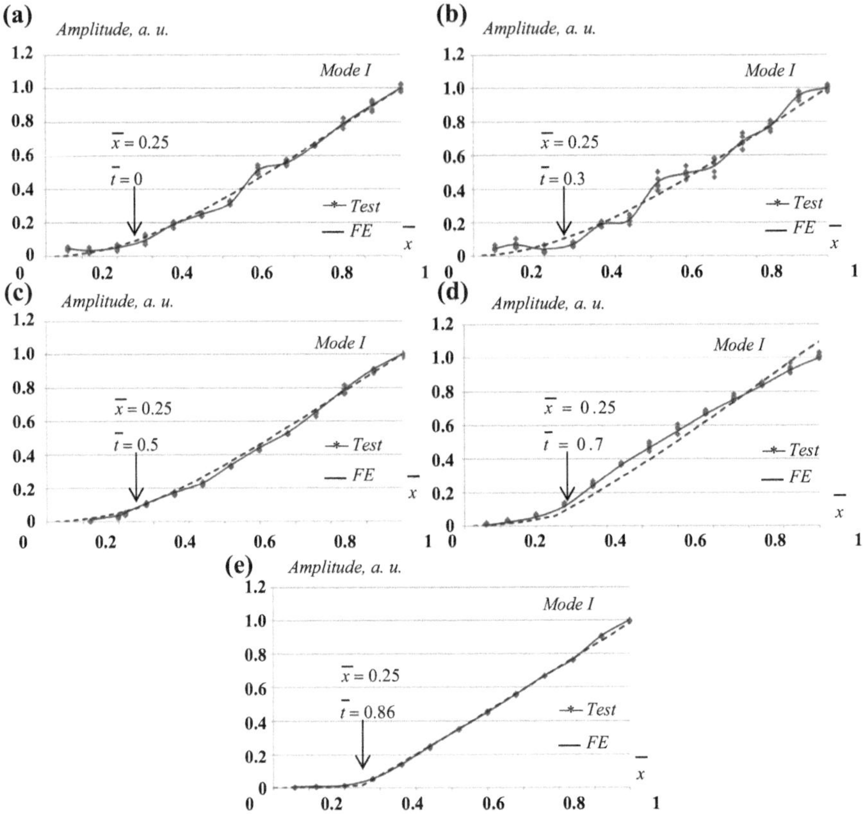

Fig. 7.13 Oscillation forms of Mode I in vertical plane of beam with notch at its location $L_c = 0.25$ for the defect size t: (**a**) 0; (**b**) 0.30; (**c**) 0.50; (**d**) 0.70; (**e**) 0.86
• Test values of oscillation amplitudes at various points; — averaged curve of test values of oscillation amplitudes; — numerical calculation using software ANSYS

horizontal face of the sample, the coordinates of which were fixed. At each point, five measurements of the oscillation amplitude, excited at the resonance frequency, were measured. The values of the sets of amplitudes for each point were averaged. Plots of the forms of transverse oscillations in the vertical plane of the model of the first three modes were made. Forms of oscillations, based on the solution of the modal problem using the FE software ANSYS were also constructed.

To compare the experimental and calculated data on the oscillation modes, the amplitudes were normalized in respect to the amplitude at a point located at the free edge of the beam. The forms of the first, second, and third oscillation modes obtained experimentally and compared with the calculated ones are shown in Figs. 7.13, 7.14, and 7.15, respectively.

The scatter of the amplitudes at various points of the oscillation forms of the full-scale beam model lies within 7.5%, which is a satisfactory value. The approximation

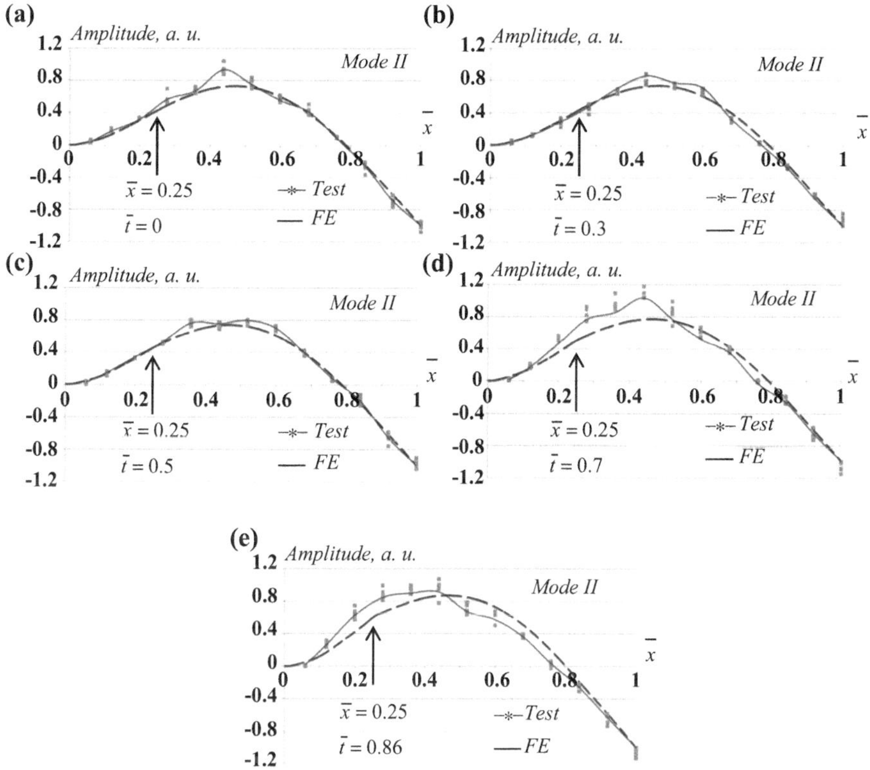

Fig. 7.14 Oscillation forms of Mode II in vertical plane of beam with notch at its location $L_c = 0.25$ for the defect size t: (**a**) 0; (**b**) 0.30; (**c**) 0.50; (**d**) 0.70; (**e**) 0.86

• Test values of oscillation amplitudes at various points; — averaged curve of test values of oscillation amplitudes; — numerical calculation using software ANSYS

of the description of oscillation forms, obtained experimentally with respect to the vibration forms, obtained by the FE modeling, depends on: (i) the accuracy of an optical sensor, (ii) the purity of the reflecting surface, (iii) the angle of inclination of the reflecting surface compared to the receiving surface of the optical sensor, (iv) the distance from the sensor to the surface of the beam, (v) the stability of the parameters of the vibro-excitatory electromagnet and the set-up as a whole, and (vi) the accuracy of excitation of the resonance frequency and the corresponding amplitudes in power to the noise threshold of the oscillations.

7.3.3.3 Comparison of Oscillation Forms, Obtained Experimentally and in FE-Modeling

A comparative analysis of the plots of the oscillation forms, constructed from the results of numerical calculations (Fig. 7.16 a, c, d) and from the data of the physical

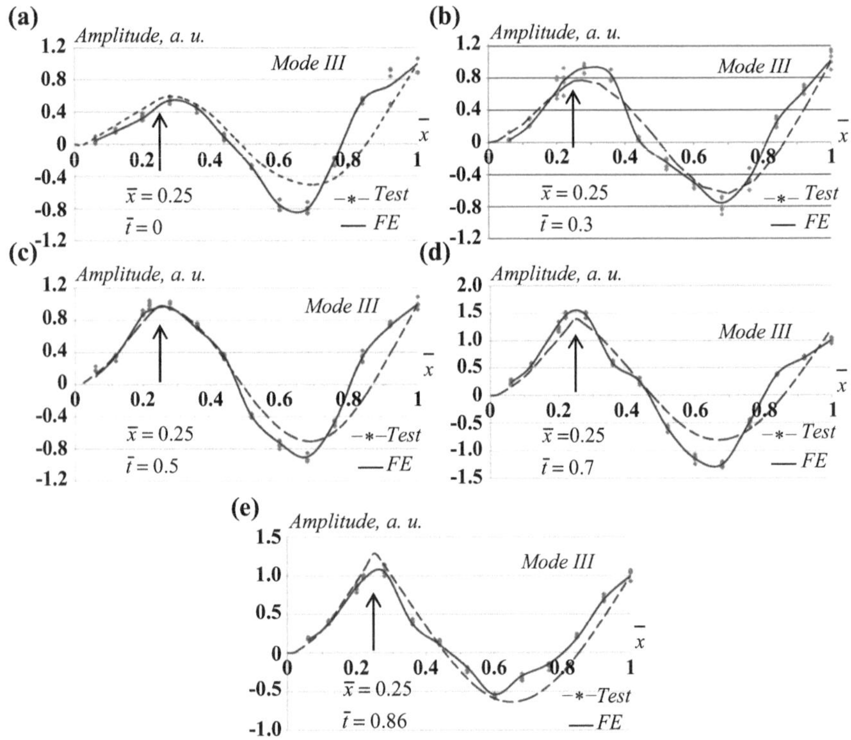

Fig. 7.15 Oscillation forms of Mode III in vertical plane of beam with notch at its location L_c= 0.25 for the defect size t: (**a**) 0; (**b**) 0.30; (**c**) 0.50; (**d**) 0.70; (**e**) 0.86

• Test values of oscillation amplitudes at various points; — averaged curve of test values of oscillation amplitudes; — numerical calculation using software ANSYS

experiment (Fig. 7.16 b, d, e) showed pronounced features, namely: kinks (bends) of the forms at a notch location $L_c = 0.25$ with a depth $t = 0.30, 0.50, 0.70, 0.86$, which were absent on the plots of oscillation modes for the intact beam. On both plots of oscillation forms, these kinks (bends) are well-defined for curves, describing the oscillations of a beam with a notch of depth $t \geq 0.50$. Their difference is based on the fact that on experimental plots for beams with a smaller depth of notch, in addition to the kinks of the curves at $L_c = 0.25$, there are several kinks at points with coordinates $L_c > 0.25$. On the plots, obtained from FE calculations, there are no such kinks. This is because when data is collected on the movement of the surface of the samples with the help of a triangulation sensor, there is a spread of the displacement parameters.

In this section, the oscillation forms of a rod with a notch are obtained for the first three natural frequencies. As can be seen, in Fig. 7.16 (d and e) the features described above (curve breaks) are much stronger than on the curves for the first and second modes of oscillations. Therefore, the coordinate of the kink of the oscillation forms of the third mode, coinciding with the location of the defect, can be accepted as the primary diagnostic sign, characterizing the location of the defect in the structural

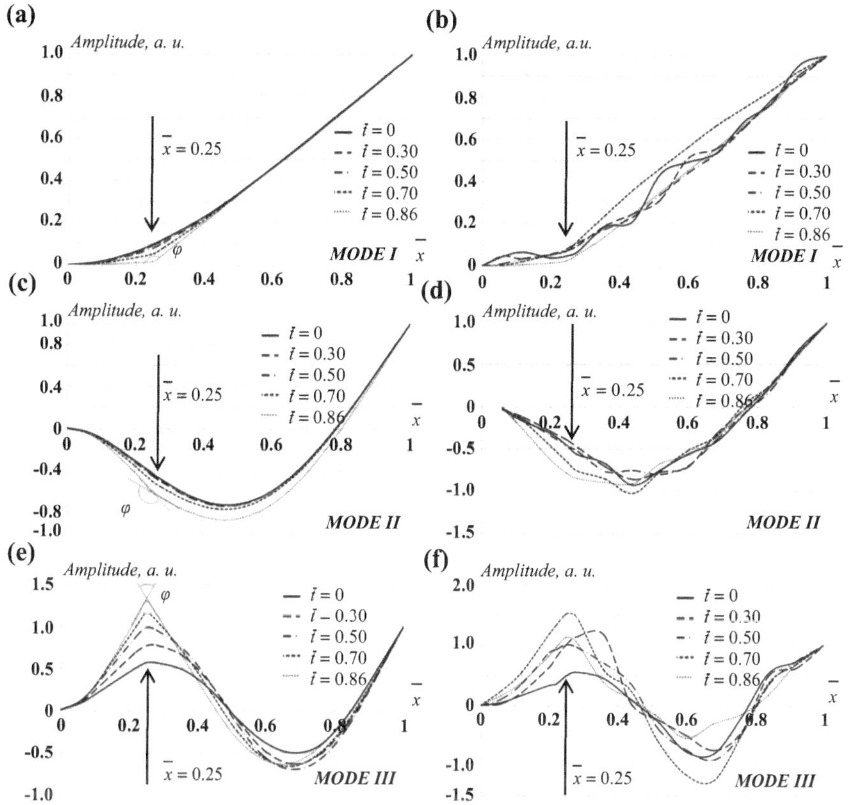

Fig. 7.16 Amplitudes of the first, second, and third modes of transverse oscillations of beam with notch (at a point separated from the pinch at a distance $L_c = 0.25$) for different depths of the notch: (**a**, **c**, **e**) correspond to numerical calculations; (**b**, **d**, **f**) correspond to full-scale tests

element. The analysis shows that the error in locating the defect using experimental data on the parameters of the oscillation form curves does not exceed 8%, so the experimental identification method can be implemented in practice.

The dynamics of the change in the defect size can be estimated by the angle φ between the tangents to the plot of the oscillation forms at the point of the beam, corresponding to the coordinate of the defect. To quantify the dynamics of the angle change with increasing notch depth, the values of the angle φ were calculated depending on the location L_c of the notch at different depths t of the cut. Based on the results of the FE calculations, the dependency curves $\phi\ (L,\ t)$were plotted for the first three oscillation modes (Fig. 7.17).

Analysis of these plots showed that the angle φ between the tangents decreases with increasing notch depth from $t - 0.3$ to $t - 0.86$ only for the first and third modes of oscillation, and this feature is clearly manifested at the point of the beam, located at a distance $L_c = 0.25$ from its pinching. This decrease in the angle φ is observed both in the calculated and experimental plots.

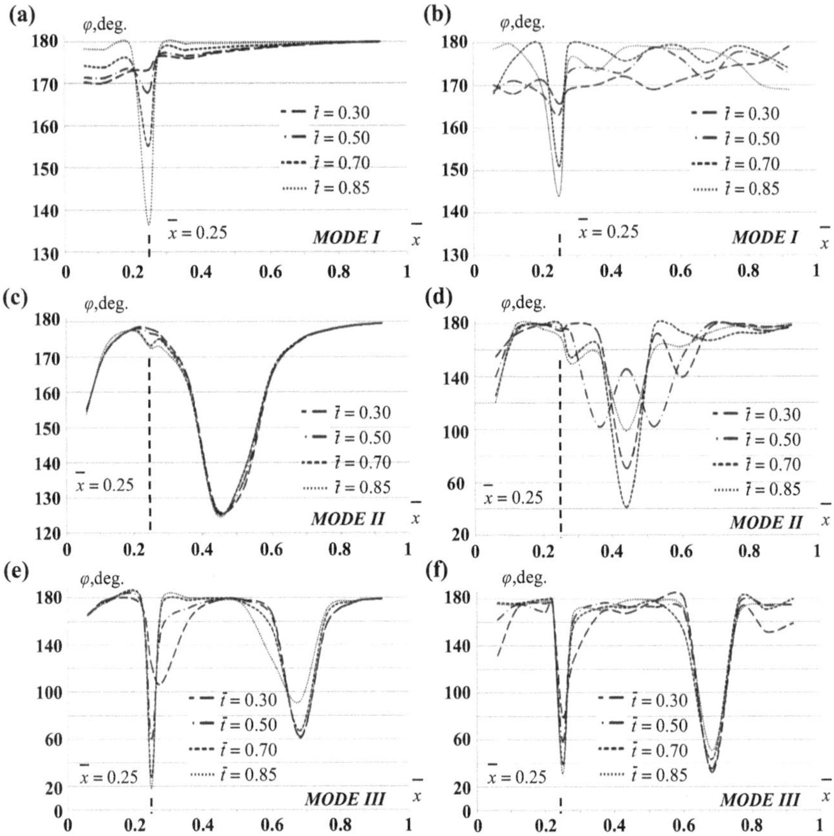

Fig. 7.17 Change in angle φ at the points between the tangents to the plots of the oscillation forms at different depths of the beam notch t, obtained in the result of FE-calculations (**a**, **c**, **e**) and experiments (**b**, **d**, **f**): (**a**, **b**)—for the first mode; (**c**, **d**)—for the second mode; (**e**, **f**)—for the third mode (location of the notch $L_c = 0.25$)

To quantify the dynamics of changes in the angles between the tangents of plot $\phi\,(L, t)$, the calculated data were processed, and the results are presented in Table 7.2. From these data, two conclusions can be made: firstly, the angle φ varies little in the range of variation of the notch depth t from 0 to 0.30, and secondly, in the range of t from 0.3 to 0.7, the angles φ, related to the first and third modes of oscillations, significantly decrease with increasing notch depth.

In the second oscillation mode, this feature is not manifested. It is also important to note that the strongest decrease of the angle φ occurs in the third mode of oscillation, and is equal to 84.1% versus 22.6% for the case of oscillations of the first mode. This is also seen from the plot of the dependences $\phi(t)$ for all three modes of oscillations shown in Fig. 7.18.

Analysis of plots of dependencies and tabular data shows that the depth of the notch of the beam can be reliably determined based on the data on the variation of the

Table 7.2 Angles φ on the graphs of oscillation forms at point with coordinate $L_c = 0.25$

	Angle ϕ between tangents, deg. $\left(\frac{\text{FEcalculation}}{\text{experiment}}\right)$			
	Relative depth of notch, t			
Number of natural oscillation mode	0.3	0.5	0.7	0.86
1	*173.2* 165.6	*167.8* 163.0	*155.1* 151.0	*136.3* 144.1
2	*178.0* 174.4	*176.8* 176.4	*173.3* 179.3	*172.6* 169.3
3	*124.1* 80.1	*60.3* 58.4	*27.8* 39.2	*18.3* 31.0

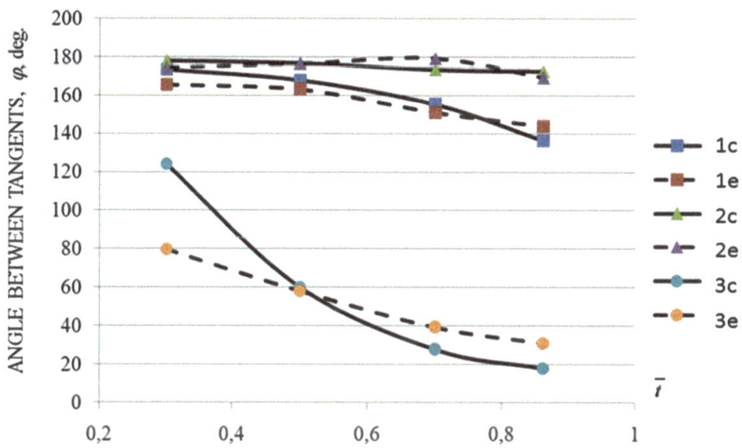

Fig. 7.18 Change in the angle φ between the tangents to the plots of the oscillation forms at the point of the beam with the coordinate $L_c = 0.25$: 1c, 2c, and 3c are the calculated values for the first, second, and third modes of vibration, respectively; 1e, 2e, and 3e are the values obtained experimentally for the first, second, third vibration modes, respectively

angle φ between the tangents to the curves of the forms of the third oscillation mode with the coordinate $L_c = 0.25$, which coincides with the defect coordinate on the beam. Here it is necessary to add that with the help of the proposed diagnostic sign, it is possible to estimate the relative depth of the notch, exceeding 20% of the height of the beam cross-section.

Figure 7.19 presents plots of curvature variation at different points for the first oscillation form of the beam model in dependence on the depth t of the notch at the cut location $L_c = 0.25$. The method for calculating the magnitude of the curvature is presented in Chap. 6. Examination of the distinctive features of the presence of a defect on the curve of the oscillation form was performed on the base of the FE calculations using software ANSYS. The plots of the curvature of the vibration shape demonstrate steep bend at the site of the defect location. The points of the amplitude vector calculation are located in the places considered in the experiment.

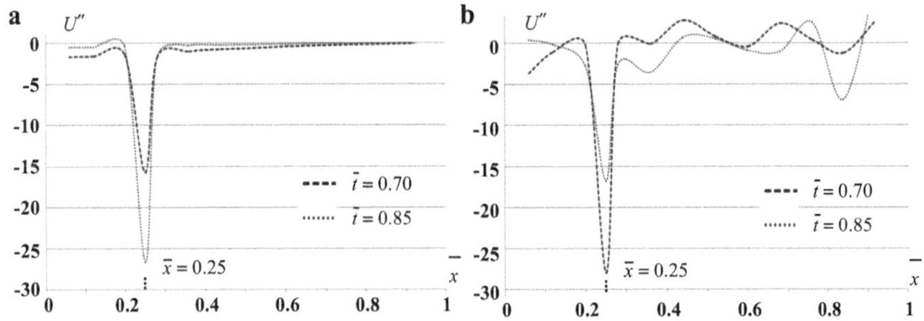

Fig. 7.19 Curvature U'' of the first oscillation form for various values of notch depth t; plots obtained in calculations (**a**) and experiments (**b**) (the notch location $L_c = 0.25$)

The plots of curves of vibration shapes obtained based on processing the experimental results, at the sizes of the beam notch $t \leq 0.5$, have a wide scatter of data. Thus, the determination of the notch location in the beam for these notches is difficult due to errors associated with measuring the parameters of the oscillation modes.

7.4 Conclusions

1. The technical capabilities, composition, and construction of the experimental sample of a multichannel information measuring system are considered as a set-up providing automated data collection on oscillatory processes and possessing an assessment of damage in the vibration diagnostics of defects in the elements of rod constructions. The software "VibroGraf" makes it possible to automate the process of measuring the vibration parameters and obtaining dynamic deformation images of the investigated rod structures. An example of the operation shows that the use of hardware allows identification of defects in the rod constructions.
2. Approbation of the calculation-experimental approach to the determination of defects in the cantilever beam construction was carried out. Analysis of the oscillations showed that the deviation of the first seven resonance frequencies of oscillations, obtained by the FE simulation, from the resonance frequencies, obtained experimentally, did not exceed 12.9%.
3. The application of the algorithm for identification of defects in beam constructions is presented. The above-mentioned example shows sufficient agreement between numerical and experimental results, and confirms the operability of the proposed identification method.

References

1. Abdollahi, A., Peco, C., Millán, D., Arroyo, M., & Arias, I. (2014). *Journal of Applied Physics, 116*(9), 093502.
2. Adams, R. D., Cawley, P., Pye, C. J., & Stone, B. J. (1978). *Journal of Mechanical Engineering Science, 20*, 93.
3. Adhikari, S., Friswell, M. I., & Inman, D. J. (2009). *Smart Materials and Structures, 18*(11), 115005.
4. *Advanced materials – manufacturing, physics, mechanics and applications*, Springer Proceedings in Physics, Ivan A. Parinov, Shun-Hsyung Chang, Vitaly Yu. Topolov (Eds.). Springer Cham, Heidelberg, New York, Dordrecht, London, *175*, 707 p. (2016).
5. *Advanced materials – physics, mechanics and applications*. Springer Proceedings in Physics, Shun-Hsyung Chang, Ivan A. Parinov, Vitaly Yu. Topolov (Eds.). Springer Cham, Heidelberg, New York, Dordrecht, London, *152*, 380 p. (2014).
6. *Advanced materials – studies and applications*, I. A. Parinov, S. H. Chang, S. Theerakulpisut (Eds.). Nova Science Publishers, New York, 527 p. (2015).
7. *Advanced materials – techniques, physics, mechanics and applications*, Springer Proceedings in Physics, Ivan A. Parinov, Shun-Hsyung Chang, Muaffaq A. Jani (Eds.). Springer Cham, Heidelberg, New York, Dordrecht, London, *193*, 637 p. (2017).
8. *Advanced nano- and piezoelectric materials and their applications*, Ivan A. Parinov (Ed.). Nova Science Publishers, New York, 250 p. (2014).
9. Akopyan, V. A. (2009). *Russian Journal of Nondestructive Testing, 45*(3), 24.
10. Akopyan, V. A., Cherpakov, A. V., Rozhkov, E. V., & Soloviev, A. N. (2012). *Testing and Diagnostics, 7*, 50.
11. Akopyan, V. A., Cherpakov, A. V., Soloviev, A. N., Kabel'kov, A. N., Shevtsov, S. N. (2010). *Izvestiya VUZov. North-Caucasus Region. Technical Sciences, 5*, 21 (In Russian).
12. Akopyan, V. A., Kabel'kov, A. N., Cherpakov, A. V. (2009). *Izvestiya VUZov. North-Caucasus Region. Technical Sciences, 5*, 89 (In Russian).
13. Akopyan, V. A., Parinov, I. A., Rozhkov, E. V., Soloviev, A. N., Shevtsov, S. N., Cherpakov, A. V. (2010). *Russian Patent for Utility Model* No. 94302, 20.05.2010 (In Russian).
14. Akopyan, V. A., Parinov, I. A., Zakharov, Y. N., Chebanenko, V. A., & Rozhkov, E. V. (2015). In I. A. Parinov, S. H. Chang, & S. Theerakulpisut (Eds.), *Advanced materials – studies and applications* (p. 417). New York: Nova Science Publishers.
15. Akopyan, V. A., Rozhkov, E. V., Cherpakov, A. V., Soloviev, A. N., Kabel'kov, A. N., & Shevtsov, S. N. (2010). *Testing and Diagnostics, 2*, 49.

S. N. Shevtsov et al., *Piezoelectric Actuators and Generators for Energy Harvesting*, Innovation and Discovery in Russian Science and Engineering,
https://doi.org/10.1007/978-3-319-75629-5

16. Akopyan, V. A., Rozhkov, E. V., Soloviev, A. N., Shevtsov, S. N., & Cherpakov, A. V. (2015). *Identification of damages in elastic constructions: Approaches, methods, analysis* (p. 74). Rostov-on-Don: Southern Federal University Press (In Russian).
17. Akopyan, V., Soloviev, A., & Cherpakov, A. (2010). In A. L. Galloway (Ed.), *Mechanical vibrations: Types, testing and analysis* (p. 147). New York: Nova Science Publishers.
18. Akopyan, V. A., Solov'ev, A. N., Cherpakov, A. V., & Shevtsov, S. N. (2013). *Russian Journal of Nondestructive Testing, 49*(10), 579.
19. Akopyan, V. A., Soloviev, A. N., Kabel'kov, A. N., Cherpakov, A. V. (2009). *Izvestiya VUZov. North-Caucasus Region. Technical Sciences, 1*, 55 (In Russian).
20. Akopyan, V. A., Soloviev, A. N., Parinov, I. A., & Shevtsov, S. N. (2010). *Definition of constants for piezoceramic materials* (p. 205). New York: Nova Science Publishers.
21. Anton, S. R. (2011). *Multifunctional piezoelectric energy harvesting concepts*. PhD Thesis, Virginia Polytechnic Institute and State University, Blacksburg. Virginia.
22. Baker, J., Roundy, S., Wright, P. (2005). In: *3rd International Energy Conversion Engineering Conference*, 170.
23. Bamnios, Y., Douka, E., & Trochidis, A. (2002). *Journal of Sound and Vibration, 256*(2), 287.
24. Belokon, A. V., & Skaliuh, A. S. (2010). *Mathematical Modeling of Irreversible Processes of Polarization*. Moscow: Fizmatlit. (in Russian).
25. Bendsoe, M. P., & Sigmund, O. (2004). *Topology optimization: Theory, methods and applications*. Berlin: Springer.
26. Biscontin, G., Morassi, A., & Wendel, P. (1998). *Journal of Vibration and Control, 4*, 237.
27. Bishop, R. E. D., & Johnson, D. C. (1960). *The mechanics of vibration*. Cambridge: Cambridge University Press.
28. Boltezar, M., Strancar, B., & Kuhelj, A. (1998). *Journal of Sound and Vibration, 211*, 729.
29. Bovsunovsky, A. P., & Surace, C. (2005). *Journal of Sound and Vibration, 288*, 865.
30. Bovsunovsky, A. P., & Matveev, V. V. (2000). *Journal of Sound and Vibration, 235*, 415.
31. Bovsunovsky, O. A. (2008). *Strength Problems, 5*, 114.
32. Bowen, C. R., Butler, R., Jervis, R., Kim, H. A., & Salo, A. I. T. (2007). *Journal of Intelligent Material Systems and Structures, 18*(1), 89.
33. Burtseva, O. V., Kosenko, E. E., Kosenko, V. V., Nefedov, V. V., Cherpakov, A. V. (2011). *Don Engineering Bulletin, 4* (In Russian).
34. Capecchi, D., & Vestroni, F. (1999). *Earthquake Engineering and Structural Dynamics, 28*, 447.
35. Cavallier, B., Berthelot, P., Nouira, H., Foltete, E., Hirsinger, L., & Ballandras, S. (2005). *IEEE Ultrasonics Symposium, 2*, 943.
36. Cawley, P., & Adams, R. D. (1979). *Journal of Strain Analysis, 14*, 49.
37. Cerri, M. N., & Vestroni, F. (2000). *Journal of Sound and Vibration, 234*, 259.
38. Chebanenko, V. A., Akopyan, V. A., & Parinov, I. A. (2015). In I. A. Parinov (Ed.), *Piezoelectrics and nanomaterials: Fundamentals, developments and applications* (p. 243). New York: Nova Science Publishers.
39. Cherpakov, A. V., Akopyan, V. A., & Soloviev, A. N. (2013). *Technical Acoustics, 13*, 1. (In Russian).
40. Cherpakov, A. V., Akopyan, V. A., Soloviev, A. N., Rozhkov, E. V., & Shevtsov, S. N. (2011). *Bulletin of Don State Technical University, 11*(3), 312. (In Russian).
41. Cherpakov, A. V., Kayumov, R. A., Kosenko, E. E., & Mukhamedova, I. Z. (2014). *Bulletin of Kazan Technological University, 17*(10), 182 (In Russian).
42. Cherpakov, A. V., Soloviev, A. N., Chakraverty, S., Gricenko, V. V., Butenko, U. I., & Bocharova, O. V. (2014). *Engineering Journal of Don, 31*(4), 29 (In Russian).
43. Cherpakov, A. V., Soloviev, A. N., Gritsenko, V. V., & Goncharov, O. U. (2016). *Defence Science Journal, 66*(1), 44.
44. Chondros, T. G., & Dimarogonas, A. D. (1980). *Journal of Sound and Vibration, 69*, 531.
45. Chondros, T. G., Dimarogonas, A. D., & Yao, J. (1998). *Journal of Sound and Vibration, 215*, 17.

46. Chondros, T. G., Dimarogonas, A. D., & Yao, J. (2001). *Journal of Sound and Vibration, 239*, 57.
47. Christides, S., & Barr, A. D. S. (1984). *International Journal of Mechanical Sciences, 26*, 639.
48. Coppotelli, G., Agneni, A., & Balis Crema, L. (2008). ISMA 2008 Conf., Leuven, Belgium, 18–20 Sept 2008.
49. Dado, M. H. F., & Shpli, O. A. (2003). *International Journal of Solids and Structures, 40*, 5389.
50. Danciger, A. Ya., Razumovskaya, O. N., Reznichenko, L. A., Sahnenko, V. P., Klevcov, A. N., Dudkina, S. I., Shilkina, L. A., Dergunova, N. V., Rybyanec, A. N. (2001, 2002). *Multicomponent systems of ferroelectric complex oxides: Physics, crystallography, technology. Design aspects of ferro-piezoelectric materials.* (p. 800). Rostov-on-Don: Rostov State University Press, *1*, *2*, (In Russian).
51. Davini, C., Gatti, F., & Morassi, A. (1993). *Meccanica Journal of the Italian Association of Theoretical and Applied Mechanics, 28*, 27.
52. Beda, P. I. (Ed.). (1978). *Defectoscopy of details at exploring of aviation techniques*. Moscow: Voenizdat. (In Russian).
53. Deng, Q., Kammoun, M., Erturk, A., & Sharma, P. (2014). *International Journal of Solids and Structures, 51*(18), 3218.
54. Dilena, M. (2003). In C. Davini & E. Viola (Eds.), *Problems in structural identification and diagnostics: General aspects and applications*. New York: Springer.
55. Dilena, M., & Morassi, A. (2004). *Journal of Sound and Vibration, 276*(1–2), 195.
56. Dirr, B. O., & Schmalhorst, B. K. (1988). Transactions of the American society of mechanical engineers, series B. *Journal of Engineering for Industry, 110*, 158.
57. Duong, L. V., Pham, M. T., Chebanenko, V. A., Solovyev, A. N., & Nguyen, C. V. (2017). *International Journal of Applied Mechanics, 9*(6), 1750084.
58. DuToit, N. E., & Wardle, B. L. (2007). *AIAA Journal, 45*(5), 11261137.
59. Dutoit, N. E., Wardle, B. L., & Kim, S. G. (2005). *Integrated Ferroelectrics, 71*(1), 121.
60. Elvin, N., & Erturk, A. (2013). *Advances in energy harvesting methods*. Heidelberg: Springer.
61. Erturk, A., Inman, D. J., & Intell, J. (2008). *Journal of Intelligent Material Systems and Structures, 19*(11), 1311.
62. Erturk, A., & Inman, D. J. (2011). *Piezoelectric energy harvesting*. New York: John Wiley and Sons, Ltd..
63. Feenstra, J., Granstrom, J., & Sodano, H. (2008). *Mechanical Systems. and Signal Processing, 22*(3), 721.
64. Parinov, I. A. (Ed.). (2012). *Ferroelectrics and superconductors: Properties and applications* (p. 287). New York: Nova Science Publishers.
65. Freund, L. B., & Herrmann, G. (1976). *Journal of Applied Mechanics, 76*, 112.
66. Gash, R., Person, M., & Weitz, B. (1983). *Institute of Mechanical Engineering, C311/88*, 163.
67. Ya, A., & Gladkov. (1989). *Technical Diagnostics and Non-destructive Testing, 4*, 14.
68. Gladwell, G. M. L., & Morassi, A. (1999). *Inverse Problems in Engineering, 7*, 215.
69. Goldfarb, M., & Jones, L. D. (1999). *Transactions-American Society of Mechanical Engineers Journal of Dynamic Systems Measurement and Control, 121*, 566.
70. Gounaris, G., & Dimarogonac, A. D. (1988). *Computer and Structures., 28*, 309.
71. Gudmundson, P. (1982). *Journal of the Mechanics and Physics of Solids, 30*, 339.
72. Hald, O. H. (1984). *Communications on Pure and Applied Mathematics, 37*, 539.
73. Han, B., Vassilaras, S., Papadias, C. B., Soman, R., Kyriakides, M. A., Onoufriou, T., & Prasad, R. (2013). *Journal of Vibration and Control, 19*(15), 2255.
74. Hearn, G., & Testa, R. B. (1991). *Journal of Structural Engineering ASCE, 117*, 3042.
75. Hofinger, M., & Leconte, P. (2004). 6th ONERA-DLR Aerospace Symp, 22–23 June 2004.
76. Hu, S., & Shen, S. (2010). *Science China Physics, Mechanics and Astronomy, 53*(8), 1497.
77. Huang, X., & Xie, Y. M. (2010). *Evolutionary topology optimization, methods and applications*. Chichester: John Wiley and Sons Ltd.

78. Imam, I., Azarro, S., Bankert, R., & Scheibel, J. (1989). *Journal of Vibration, Acoustics, Stress and Reliability in Design, 3*, 241.
79. Indenbom, V. L., Loginov, E. B., & Osipov, M. A. (1981). *Kristallografiya, 26*(6), 1157.
80. Johnson, W. (1982). *Self-tuning regulators for multicyclic control of helicopter vibration*, NASA Technical Report.
81. Kerr, A. D., & Alexander, H. (1968). *Acta Mechanica, 6*(2), 180.
82. Kessler, C. (2011). *CEAS Aeronautical Journal, 1*(1–4), 23.
83. Kloepel, V. (2006). International Conference Cooperation with the New EU Member States in Aeronautics Research, Berlin, May 15th, 2006.
84. Kogan, S. M. (1963). *Solid State Physics, 5*(10), 2829.
85. Kosenko, E. E., Kosenko, V. V., & Cherpakov, A. V. (2013). *Engineering Journal of Don, 27* (4), 271 (In Russian).
86. Kosenko, E. E., Kosenko, V. V., & Cherpakov, A. V. (2013). *Engineering Journal of Don, 27* (4), 272 (In Russian).
87. Krasnoshchekov, A. A., Sobol, B. V., Solov'ev, A. N., & Cherpakov, A. V. (2011). *Russian Journal of Nondestructive Testing, 47*(6), 412.
88. Krawczuk, M., & Ostachowicz, W. M. (1992). *Archives of Applied Mechanics, 62*, 463.
89. Kube, R., Schimke, D., & Janker, P. (2000). RTO AVT symposium on active control technology for enhanced performance operational capabilities of military aircraft, land vehicles and sea vehicles, Braunschweig, Germany, 8–11 May 2000.
90. Kumar, D., & Cesnik, C. E. S. (2013). AHS 69th Annual Forum, Phoenix, Arizona, 21–23 May 2013.
91. Leighton, W., & Nehari, Z. (1958). *Transactions of the American Mathematical Society, 89*, 325.
92. Liang, R. Y., Hu, J., & Choy, F. (1992). *Journal of Mathematical Analysis and Applications, 118*, 384.
93. Liang, R. Y., Hu, J., & Choy, F. (1992). *Journal of Mathematical Analysis and Applications, 118*, 1469.
94. Liang, X., Hu, S., & Shen, S. (2013). *Journal of Applied Mechanics, 80*(4), 044502.
95. Liao, Y., Sodano, H. A. (2009). *20*(5), 505.
96. Liu, L., Friedmann, P. P., Kim, I., & Bernstein, D. S. (2006). AHS 62nd Annual Forum, Phoenix, AZ.
97. Liu, Y., Tian, G., Wang, Y., Lin, J., Zhang, Q., & Hofmann, H. F. (2009). *Journal of Intelligent Material Systems and Structures, 20*(5), 575585.
98. Ma, W., & Cross, L. E. (2003). *Applied Physics Letters, 82*(19), 3293.
99. Mainz, H., van der Wall, B. G., Leconte, F., Ternoy, F., & des Roshettes, H. M. (2005). 31st European Rotorcraft Forum, Florence, Italy, 13–15 September 2005.
100. Majdoub, M. S., Sharma, P., & Cagin, T. (2008). *Physical Review B, 77*, 125424.
101. Makhutov, N. A. (2005). *Construction strength, resource and technogenic safety*. Novosibirsk: Nauka. (In Russian).
102. Maranganti, R., Sharma, N. D., & Sharma, P. (2006). *Physical Review B, 74*(1), 014110.
103. Mashkevich, V. S., & Tolpygo, K. B. (1957). *Soviet Physics - JETP, 4*, 455.
104. Matveev, V. V. (1997). *Strength Problems, 6*, 5.
105. Matveev, V. V., & Bovsunovsky, A. P. (1999). *Strength Problems, 4*, 19.
106. Matveev, V. V., & Bovsunovsky, A. P. (2000). *Strength Problems, 5*, 44.
107. Matveev, V. V., & Bovsunovsky, A. P. (2002). *Journal of Sound and Vibration, 249*(1), 23.
108. Maucher, C. K. et al. (2007). 33rd European Rotorcraft Forum, Kazan, Russia.
109. Milov, A. B., Nevsky, Y. N., Strakhov, G. A., & Ukhov, Y. G. (1983). *Issues of dynamics and strength* (Vol. 43, p. 97). Riga: Zinatne. (In Russian).
110. Mindlin, R. D. *Problems of continuum mechanics* In: Soc. Industrial and Appl. Math., Philadelphia, J. R. M., Ed., 282 (1961).
111. Moosad, K. P. B., Krishnakumar, P., Chandrashekar, G., & Vishnubhatla, R. M. R. (2007). *Applied Acoustics, 68*(10), 1063.
112. Morales, R. M., & Turner, M. C. (2013). AHS 69th Annual Forum, Phoenix, Arizona, 21–23 May 2013.

113. Morassi, A. (1993). *Journal of Mathematical Analysis and Applications, 119*, 1798.
114. Morassi, A. (1997). *Inverse Problems in Engineering, 4*, 231.
115. Morassi, A. (2001). *Journal of Sound and Vibration, 242*, 577.
116. A. Morassi, In: Problems in structural identification and diagnostics: General aspects and applications, C. Davini, E. Viola, Eds. Springer, New York, 163 (2003).
117. Morassi, A., & Dilena, M. (2002). *Inverse Problems in Engineering, 10*, 183.
118. Morassi, A., & Rollo, M. (2001). *Journal of Vibration and Control, 7*, 729.
119. Morassi, A., & Rovere, N. (1997). *Journal of Mathematical Analysis and Applications, 123*, 422.
120. Parinov, I. A. (Ed.). (2013). *Nano- and piezoelectric technologies, materials and devices* (p. 261). New York: Nova Science Publishers.
121. Narkis, Y. (1994). *Journal of Sound and Vibration, 172*, 549.
122. Natke, H. G., & Cempel, C. (1991). *Journal of Mechanical Systems and Systems Processing, 5*, 345.
123. Nechibvute, A., Chawanda, A., & Luhanga, P. (2012). *ISRN Materials Science, 2012*, 921361.
124. Nelson, H. D., & Nataraj, C. (1986). *Transactions of the American Society of Mechanical Engineers, Series B. Journal of Engineering for Industry, 108*, 52.
125. *Non-destructive testing and diagnostics: Handbook*, V. V. Klyuev etc., (Ed.). Moscow: Mashinostroenie, (1995) (In Russian).
126. Ostachowicz, W. M., & Krawczuk, M. (1990). *Computers and Structures, 36*, 245.
127. Ostachowicz, W. M., & Krawczuk, M. (1991). *Journal of Sound and Vibration, 150*, 191.
128. Pandey, A. K., Biswas, M., & Samman, M. M. (1991). *Journal of Sound and Vibration, 145*, 321.
129. Papatheou, E., Mottershead, J. E., & Cooper, J. E. (2013). Int. Conf. Structural Engineering Dynamics ICEDyn 2013, Lisbon, Portugal, 17–19 June 2013.
130. Parinov, I. A. (2012). *Microstructure and properties of high-temperature superconductors* (Second ed. p. 779). Heidelberg, New York, Dordrecht, London: Springer.
131. Paris, P., & Si, J. (1968). *Applied issues of fracture toughness* (p. 64). Moscow: Mir. (In Russian).
132. Parton, V. Z., & Morozov, E. M. (1985). *Mechanics of elasto-plastic fracture*. Moscow: Nauka. (In Russian).
133. Pavelko, I. V., Pavelko, V. P. (2002). *Scientific Proceedings of Riga Technical University. Ser. 6. Transport and Engineering. Mechanics*. RTU Press, Riga, 7, 159.
134. Pavelko, V. P., Shakhmansky, G. V. (1971). *Proc. RIICA*, Riga, *191*, 87. (In Russian).
135. Parinov, I. A., & Hsyung-Chang, S. (Eds.). (2013). *Physics and mechanics of new materials and their applications* (p. 444). New York: Nova Science Publishers.
136. Parinov, I. A. (Ed.). (2010). *Piezoceramic materials and devices* (p. 335). New York: Nova Science Publishers.
137. Parinov, I. A. (Ed.). (2012). *Piezoelectric materials and devices* (p. 328). New York: Nova Science Publishers.
138. Parinov, I. A. (Ed.). (2015). *Piezoelectrics and nanomaterials: Fundamentals, developments and applications* (p. 283). New York: Nova Science Publishers.
139. Parinov, I. A. (Ed.). (2012). *Piezoelectrics and related materials: Investigations and applications* (p. 306). New York: Nova Science Publishers.
140. Postnov, V. A. (2000). *Izvestiya of RAS. Mechanics of Solids, 6*, 155.
141. Prechtl, E. F., & Hall, S. R. (1997). *Proc. SPIE* 3041, Smart Structures and Materials 1997: Smart Structures and Integrated Systems, 158.
142. Rizos, P. F., Aspragathos, N., & Dimarogonas, A. D. (1990). *Journal of Sound and Vibration, 143*, 381.
143. Roundy, S., & Wright, P. K. (2004). *Smart Materials and Structures, 13*(5), 1131.
144. Roundy, S., & Wright, P. K. (2004). *Smart Materials and Structures, 13*(5), 1135.
145. Ruotolo, R., Surace, C., Crespo, P., & Storer, D. (1996). *Computers and Structures, 61*(6), 1057.
146. Rytter, A. (1999). *Vibration based inspection of civil engineering structures*. PhD Thesis, University of Aalborg (Denmark), 193 p.
147. Saavedra, P. N., & Cuitino, L. A. (2001). *Computers and Structures, 79*, 1451.

148. Sahin, E., & Dost, S. (1988). *International Journal of Engineering Science, 26*(12), 1231.
149. Shen, J., Chopra, I., & Johnson, W. (2003). AHS 59th Annual Forum, Phoenix, Arizona, 6–8 May 2003.
150. Shen, J., Yang, M., & Chopra, I. (2004). 45th AIAA/ASME/ASCE/AHS/ASC Structures, Structural Dynamics & Materials Conf., Palm Springs, Ca, 19–22 Apr 2004.
151. Shen, M.-H. H., & Pierre, C. (1990). *Journal of Sound and Vibration, 138*, 115.
152. Shen, M.-H. H., & Taylor, J. E. (1991). *Journal of Sound and Vibration, 150*, 457.
153. Shevtsov, S., Akopyan, V., Rozhkov, E. (2011). *Proc. of the 5th Intern. Symp. on Defect and Material Mechanics. Jule 2011. Sevilia, Spain*, 92.
154. Shevtsov, S., Akopyan, V., Rozhkov, E., Chebanenko, V., Yang, C. C., Jenny Lee, C. Y., & Kuo, C. X. (2016). In: *Advanced materials – manufacturing, physics, mechanics and applications*, Springer Proceedings in Physics, Ivan A. Parinov, Shun-Hsyung, Vitaly Yu. Topolov (Eds.), (p. 534). Heidelberg, New York, Dordrecht, London:Springer Cham, 175.
155. Shevtsov, S., Flek, M., & Zhilyaev, I. (2013). In I. A. Parinov & S.-H. Chang (Eds.), *Physics and mechanics of new materials and their applications* (p. 259). New York: Nova Science Publishers.
156. Shevtsov, S., Zhilyaev, I., & Axenov, V. (2013). IASTED Int. Conf. Modelling, Identification and Control, Innsbruck, Austria, 11–13 February 2013.
157. Soloviev, A. N., Chebanenko, V. A., & Parinov, I. A. (2018). In H. Altenbach, E. Carrera, & G. Kulikov (Eds.), *Analysis and modelling of advanced structures and smart systems, series: Advanced structured materials* (p. 227). Singapore: Springer Nature.
158. Soloviev, A. N., Chebanenko, V. A., Zakharov, Yu. N., Rozhkov, E. V., Parinov, I. A., & Gupta, V. K. (2017). In: *Advanced materials — techniques, physics, mechanics and applications*, Springer Proceedings in Physics, Ivan A. Parinov, Shun-Hsyung Chang, Muaffaq A. Jani, (Eds.). Heidelberg, New York, Dordrecht, London:Springer Cham, *193*, 485.
159. Soloviev, A. N., Cherpakov, A. V., & Parinov, I. A. (2016). In: *Proceedings of the 2015 international conference on physics and mechanics of new materials and their applications, devoted to 100-th Anniversary of the Southern Federal University*, Ivan A. Parinov, Shun-Hsyung Chang, Vitaly Yu. Topolov (Eds.). (p. 515). New York: Nova Science Publishers.
160. Soloviev, A. N., Duong, L. V., Akopyan, V. A., Rozhkov, E. V., & Chebanenko, V. A. (2016). *DSTU Vestnik, 1*(84), 19. (in Russian).
161. Soloviev, A. N., Parinov, I. A., Cherpakov, A. V., Chebanenko, V. A., Rozhkov, E. V., & Duong, L. V. (2018). In: *Advances in structuran integrity. Proceedings of SICE 2016, series: Mechanical engineering*, R. V. Prakash, V. Jayaram, & A. Saxena (Eds.), Singapore: Springer Nature, 291.
162. Soloviev, A. N., Parinov, I. A., Duong, L. V., Yang, C. C., Chang, S. H., & Lee, J. C. Y. (2013). In I. A. Parinov, & S.-H. Chang (Eds.), *Physics and mechanics of new materials and their applications* (p. 335). New York: Nova Science Publishers.
163. Soloviev, A. N., & Vatulyan, A. O. (2011). In I. A. Parinov (Ed.), *Piezoceramic materials and devices* (p. 1). New York: Nova Science Publishers.
164. Solovyev, A. N., & Duong, L. V. (2016). *Journal of Applied Mechanics, 8*(3), 1650029.
165. Straub, F., Anand, V. R., Birxhette, T., & Lau, B. H. (2009). 35th European Rotorcraft Forum, Hamburg, Germany, 22–25 September 2009.
166. Sundermeyer, J. N., & Weaver, R. L. (1995). *Journal of Sound and Vibration, 183*, 857.
167. Timoshenko, S. P. (1967). *Oscillations in engineering*. Moscow: Nauka. (In Russian).
168. Tolpygo, K. B. (1962). *Solid State Physics, 4*, 1765.
169. Tsyfansky, S. L., & Beresnevich, V. I. (1998). *Journal of Sound and Vibration, 213*(1), 159.
170. Tsyfansky, S. L., & Beresnevich, V. I. (2000). *Journal of Sound and Vibration, 236*(1), 49.
171. Tsyfanskii, S. L., Beresnevich, V. I., & Lushnikov, B. V. (2008). *Non-linear vibro-diagnostics of machines and mechanisms*. Riga: Zinatne. (In Russian).
172. Tsyfanskii, S. L., Magone, M. A., & Ozhiganov, V. M. (1985). *Russian Journal of Non-destructive Testing, 21*(3), 77.

173. Tsyfanskii, S. L., Ozhiganov, V. M., Milov, A. B., & Nevsky, Y. N. (1981). *Issues of dynamics and strength* (Vol. 39, p. 3). Riga: Zinatne. (In Russian).
174. Vatulyan, A. O. (2007). *Reverse problems in mechanics of deformable solids*. Moscow: Fizmatlit (In Russian).
175. Vatulyan, A. O. (2010). *Applied Mathematics and Mechanics, 74*(6), 909.
176. Vatulyan, A. O., & Soloviev, A. N. (2009). *Direct and reverse problems for homogeneius and heterogeneous elastic and electro-elastic solids*. Rostov-on-Don: Southern Federal University Press (In Russian).
177. Vestroni, F., & Capecchi, D. (1996). *Journal of Vibration and Control, 2*, 69.
178. Vestroni, F., & Capecchi, D. (2000). *Journal of Engineering Mechanics, 126*(7), 761.
179. Viswamurthy, S. R., & Ganguli, R. (2008). *Journal of Vibration and Control, 14*(8), 1175.
180. Wang, J., Shi, Z., & Han, Z. (2013). *Journal of Intelligent Material Systems and Structures, 24*(13), 1626.
181. Wu, Q. L. (1994). *Journal of Sound and Vibration, 173*, 279.
182. Yu, S., He, S., & Li, W. (2010). *Journal of Mechanics of Materials and Structures, 5*(3), 427.
183. Yuen, M. M. F. (1985). *Journal of Sound and Vibration, 103*, 301.
184. Zhao, S., & Erturk, A. (2014). *Sensors and Actuators A: Physical, 214*, 58.
185. Zheludev, I. S. (1966). *Czechoslovak Journal of Physics, 16*(5), 368.

Index

S. N. Shevtsov et al., *Piezoelectric Actuators and Generators for Energy Harvesting*, Innovation and Discovery in Russian Science and Engineering,
https://doi.org/10.1007/978-3-319-75629-5

FSC
www.fsc.org
MIX
Papier | Fördert gute Waldnutzung
FSC® C083411

Kolibri 360 GmbH
Ferdinand-Jühlke-Straße 7
99095 Erfurt, Deutschland
produktsicherheit@kolibri360.de